Practice and Innovations in Sustainable Transport

Practice and Innovations in Sustainable Transport

Special Issue Editors

Tariq Muneer
Mehreen Saleem Gul
Eulalia Jadraque Gago

MDPI • Basel • Beijing • Wuhan • Barcelona • Belgrade • Manchester • Tokyo • Cluj • Tianjin

Special Issue Editors

Tariq Muneer
School of Engineering and
the Built Environment,
Edinburgh Napier University
UK

Mehreen Saleem Gul
School of Energy, Geoscience,
Infrastructure and Society,
Heriot-Watt University
UK

Eulalia Jadraque Gago
School of Civil Engineering,
University of Granada
Spain

Editorial Office
MDPI
St. Alban-Anlage 66
4052 Basel, Switzerland

This is a reprint of articles from the Special Issue published online in the open access journal *Energies* (ISSN 1996-1073) (available at: https://www.mdpi.com/journal/energies/special_issues/practice_and_innovations_in_sustainable_transport).

For citation purposes, cite each article independently as indicated on the article page online and as indicated below:

LastName, A.A.; LastName, B.B.; LastName, C.C. Article Title. *Journal Name* **Year**, *Article Number*, Page Range.

ISBN 978-3-03928-548-8 (Pbk)
ISBN 978-3-03928-549-5 (PDF)

© 2020 by the authors. Articles in this book are Open Access and distributed under the Creative Commons Attribution (CC BY) license, which allows users to download, copy and build upon published articles, as long as the author and publisher are properly credited, which ensures maximum dissemination and a wider impact of our publications.

The book as a whole is distributed by MDPI under the terms and conditions of the Creative Commons license CC BY-NC-ND.

Contents

About the Special Issue Editors vii

Preface to "Practice and Innovations in Sustainable Transport" ix

Colin Cochrane, Tariq Muneer and Bashabi Fraser
Design of an Electrically Powered Rickshaw, for Use in India
Reprinted from: *Energies* **2019**, *12*, 3346, doi:10.3390/en12173346 1

Ross Milligan, Saioa Etxebarria, Tariq Muneer and Eulalia Jadraque Gago
Driven Performance of Electric Vehicles in Edinburgh and Its Environs
Reprinted from: *Energies* **2019**, *12*, 3074, doi:10.3390/en12163074 23

Jing Hou, He He, Yan Yang, Tian Gao and Yifan Zhang
A Variational Bayesian and Huber-Based Robust Square Root Cubature Kalman Filter for Lithium-Ion Battery State of Charge Estimation
Reprinted from: *Energies* **2019**, *12*, 1717, doi:10.3390/en12091717 45

Fuwu Yan, Jingyuan Li, Changqing Du, Chendong Zhao, Wei Zhang and Yun Zhang
A Coupled-Inductor DC-DC Converter with Input Current Ripple Minimization for Fuel Cell Vehicles
Reprinted from: *Energies* **2019**, *12*, 1689, doi:10.3390/en12091689 69

Jean-Michel Clairand, Paulo Guerra-Terán, Xavier Serrano-Guerrero, Mario Gonzalez-Rodriguez and Guillermo Escrivá-Escrivá
Electric Vehicles for Public Transportation in Power Systems: A Review of Methodologies
Reprinted from: *Energies* **2019**, *12*, 3114, doi:10.3390/en12163114 85

About the Special Issue Editors

Tariq Muneer is a Professor of Energy Engineering at Napier University, Edinburgh, currently chairing an active group engaged in research on sustainable energy, which includes sustainable transport. Professor Muneer is an international authority on the subject of solar energy and its use in buildings with over 35 years' experience. He is the author of over 215 technical articles, most of which have been outlined in his research monographs.

Mehreen Gul is an Assistant Professor in Architectural Engineering at the School of Energy, Geoscience, Infrastructure and Society at Heriot-Watt University. Mehreen's research experience is in environmental and engineering issues associated with renewable and low carbon technologies and sustainable buildings. She has developed her own photovoltaics (PVs) research lab in her school, allowing students at all levels to conduct experimental research. Mehreen has published over 34 peer-reviewed articles and practitioner reports in high-impact and international peer reviewed journals. She is a named contributor on two Chartered Institution of Building Services Engineers (CIBSE) Guides: Guide J, Weather, Solar and Illuminance data (2002) and Guide A, Environmental Design (2015). Mehreen was awarded the CIBSE Napier Shaw Bronze Medal for best paper on an entirely new approach for estimating solar diffuse irradiance.

Eulalia Jadraque Gago is a Professor at the University of Granada in the Department of Civil Engineering Construction and Engineering Projects. Civil Engineering, M.Eng., Ph.D. with 23 international publications and 4 chapters in books of prestigious publishers, as CIBSE Guide A: Environmental design. Eulalia has over 15 international conferences and continually participates in investigation projects and contracts acting as researcher. Eulalia has directed 1 thesis and has had management charges at the University. Eulalia has been invited as researcher and professor at different universities. Since 2013, she has been a member of the Word Society of Sustainable Energy Technologies.

Preface to "Practice and Innovations in Sustainable Transport"

The issue of climate change has been discussed within the scientific community as well as in popular media to such an extent that it has become a priori to almost all discussions related to the sustainable use of energy.

For developed economies of Western Europe, transport-related global greenhouse gas (GHG) emissions are beginning to stabilize, but for the world as a whole, these emissions are rapidly rising. The present 23% share of CO_2 emissions for global transport is set to rise. Road transport is the chief contributor and is responsible for 20% of the total GHG emissions. Marine and air transport jointly contribute almost equally to the remainder. The bulk of the emissions problem lies with road transport.

As a follow-up to the Paris Agreement, which determined that for the globe to halt within a 2 °C average temperature increase, the transport sector needs to be decarbonized.

The United Nations estimates that 60% of the world's population will be living in urban areas by 2030. Cities account for 2% of the world's area and for 75% of the world's energy consumption. For over a century, the automobile has offered affordable freedom of movement within urban areas. Global registrations jumped from 980 million units in 2009 to over 1.2 billion in 2018. The world population exceeded 7 billion in 2012, and every seventh person now owns a vehicle, which, in all likelihood, is powered by an internal combustion engine (ICE). Worldwide, 18 million barrels of oil are consumed each day by the automobile sector. Annually, the vehicles emit 2.7 billion tonnes of CO_2.

The Nordic region, constituting of Denmark, Finland, Iceland, Norway, and Sweden, has taken a lead on electric vehicles (EVs) with the launch of its Nordic Electric Vehicle Outlook (NEVO2018) report. The EVs offer an elegant solution toward a significant reduction in GHG and curbside pollution, provided that the electricity for charging the vehicles is obtained from renewable means.

This Issue of *Energies* aims to address the challenge of emissions reduction from the transport sector and the potential solutions that may be available in the near and not-so-distant future.

Tariq Muneer, Mehreen Saleem Gul, Eulalia Jadraque Gago
Special Issue Editors

Article

Design of an Electrically Powered Rickshaw, for Use in India

Colin Cochrane *, Tariq Muneer and Bashabi Fraser

School of Engineering and Built Environment, Edinburgh Napier University, Edinburgh EH10 5DT, UK
* Correspondence: 40270469@napier.ac.uk

Received: 21 July 2019; Accepted: 29 August 2019; Published: 30 August 2019

Abstract: The main aim of this article is to present research findings related to the design of an electric rickshaw for use in Kolkata, India, identifying weaknesses in the current cycle rickshaw, developing a design solution to address problems found, and exploring the possibility of utilising solar power. Through research and testing it was found there were many design issues, concerns with health and safety, comfort, and ergonomics. All problems found were addressed by implementing design upgrades. The testing of the current cycle rickshaw identified the power and energy required to implement an electric drive system, where a 500 W 24 V DC motor and a battery capacity of 220 ampere hours was used in conjunction with a pedal assist system to provide a range of up to 52 km. A conceptual prototype was developed to prove the successful application of a pedal assist system, which was established as a viable option for design. The design has been critically evaluated and the relevant issues discussed.

Keywords: ssustainable transport; battery powered vehicle; electric propulsion

1. Introduction

1.1. Historical Background

Today, the rickshaw represents over a century of rich Indian culture. Originally brought to the country from China, the rickshaw was a hand pulled vehicle used by traders for transportation of goods. In 1914, the traders asked the governing body for permission to allow the transportation of people as passengers. The request was granted, and rickshaws became a mode of transport for the elite, although now it is used by people of all classes. Since its introduction in 1880, the rickshaw has advanced and evolved into different categories of vehicle: Hand-pulled, cycled, and motorised. Cycle rickshaws began to appear in India around 1930; as the price of bicycles began to fall, the popularity of cycle rickshaws began to rise, and by 1935 the cycle rickshaw outnumbered its hand-pulled counterpart [1]. The auto rickshaw was introduced in 1957 [2] powered by a 250 cc, 8 horsepower, 2 stroke, single-cylinder petrol engine; today, many auto rickshaws are liquified petroleum gas (LPG) powered. The auto rickshaw's performance in terms of top speed outclasses both the hand-pulled and cycle rickshaw and is used in a different manner to them. The auto rickshaw is typically used on more open and faster roads, whereas the others are used in narrower and more congested urban environments. It could be argued that the auto rickshaw is not a competitor for the same type of business as the others. All three forms of rickshaw are still in use to this day, compounding their heritage, tradition, and role within the Indian society.

The cycle rickshaw design has not developed a great deal since its introduction and has fallen behind the times in terms of modern-day bicycle design. Bicycles have advanced to a stage where high-grade lightweight materials and efficient electric motors can make the cyclists ride almost effortless, a vast contrast from the laborious task facing the rickshaw drivers. Cycle rickshaws are heavy and uncomfortable and are therefore exhausting for the drivers to cycle; they are ergonomically inept and often found to be in a poor working condition, only adding to the driver's exertion.

1.2. Population and Environment

India has a population of over 1.2 billion people, and the population growth rate for urban areas over the decade between the last two censuses 2001–2011 was 31.08%, showing that cities in India are becoming increasingly crowded. Kolkata, the capital of the state of West Bengal, has a population of over 14 million (urban area) meaning it has a population density of 24,000 people per square kilometre; the density of population gives an insight into the congestion the cities' roads and transport networks have to contend with. The residential streets are frequently narrow, crowded with pedestrians, and often have market stalls protruding onto the roadside; this hinders the ability of larger conventional cars and buses to navigate effectively though them; therefore, smaller more manoeuvrable vehicles, such as the rickshaws, manage considerably better [3].

1.3. Driver Background

In a culture of separate classes, cycle rickshaw drivers are regarded poorly, less respected than auto rickshaw drivers. If the drivers were to drive an E-Rickshaw, they would be seen more positively [4]. The deprived and exploited sections of society are not explicitly recognized in policy documents by the government and very little attention has been paid in humanising the livelihood of the rickshaw pullers [2] This gives an alarming understanding of how the rickshaw drivers are seen by the community they serve. It shows that little help or thought is given to them.

Drivers are prone to long-term fatigue injuries and this leads to them becoming unable to work and therefore seen as lazy among members of their community. The drivers' physical efforts result in injury [5].

1.4. Climate of India

The climate of Kolkata consists of temperatures that can reach as high as 36 °C in April and May, with average monthly temperatures ranging from 19.5 °C in January to 30.4 °C in May [6]. With rainfall of over 300 mm per month from June to September reaching 375 mm in July [6], this gives an understanding of how these extreme conditions will bring many issues for the driver, from poor road conditions to wet clothing becoming uncomfortable and heavy. Ultraviolet radiation can cause many health conditions, from cataracts to skin cancer. The UV index in India can reach above 16 in the monsoon season and is in the high to extreme range (8+) for nine months of the year [7]. Precautions recommended for high range are to wear a hat, sunglasses, sunscreen (SPF 15+), umbrella, and stay indoors from 10:00 to 16:00; the extreme range precautions advise minimum outdoor activity in addition to all the previous precautions [7]. This shows the issues the drivers have to overcome due to weather and the degree of health risks facing the drivers both short and long term.

2. Interview Data

2.1. Driver Interviews

2.1.1. Interview Design

Firstly, interview questions for the drivers were developed, for which different subject areas had to be addressed: The personal and physical circumstances of the driver, the personal opinion of the driver in relation to the rickshaw, and role of the job and rickshaw usage; an interview combining these questions gave the project an insight into the social, medical, and engineering factors that the drivers face by the occupation.

The interviews of drivers consisted of 10 questions, and 25 drivers in total were interviewed. The questions asked were set with the objective to find out the typical daily demand of the rickshaw as a vehicle, the basic profile of the driver, and his needs and wants. It is worth noting that all interview participants were male. Interview aims and limitations can be found in Appendix A.

2.1.2. Rickshaw Usage

Firstly, regarding the rickshaw performance-wise, it was found that the rickshaw was in use for, on average, 11.36 h, although not operated during the full 11.36 h as time would be spent queuing for a passenger. The average daily number of journeys was 28.2, up to a maximum of around 4 km, although a journey of between 1 and 2 km is more typical; therefore, a reasonable approximation of the daily distance could be calculated as 42 km.

The rickshaws are fitted with a bench-style seat and are capable of carrying two adults as well as the driver, although 40% of drivers claimed they would carry two adults and a child—this should be disregarded as it was not typical and was only done rarely.

2.1.3. Driver Profile

The profile of a typical rickshaw driver was derived by firstly identifying the average age. The answers were to be given in age ranges and it was found that 48% of the driver interviewed were aged between 31–40 years old, as seen in Figure 1. The average driver would work 6.4 days a week and would earn 334 rupees daily.

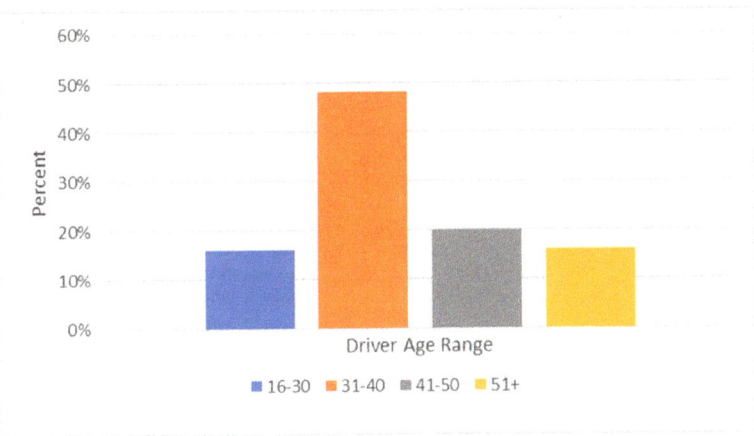

Figure 1. Driver age range.

2.1.4. Medical Issues

In total, 48% of the drivers reported injury and illness due to their occupation. These ranged from injury from collisions to breathing troubles believed to be linked to the pollution they are breathing in on a daily basis. Of the 48% of drivers reporting injury and illness, two-thirds complained of muscle soreness and aches and pains, typically in the legs and lower back. Further, 42% complained of injuries that occurred from collisions with other vehicles; one driver complained of chest pain and breathing problems (see Figure 2).

2.1.5. Driver Opinion

The drivers were asked what they would change about their rickshaw if they could. It was found that 64% wanted the implementation of an electric motor to help them power the vehicle, and when asked if an electric motor would improve their rickshaw and overall occupation, the consensus was unanimous in favour of the implementation of the electric motor. Other improvements wanted by the drivers included better shade from the weather, improved gearing, and better seating.

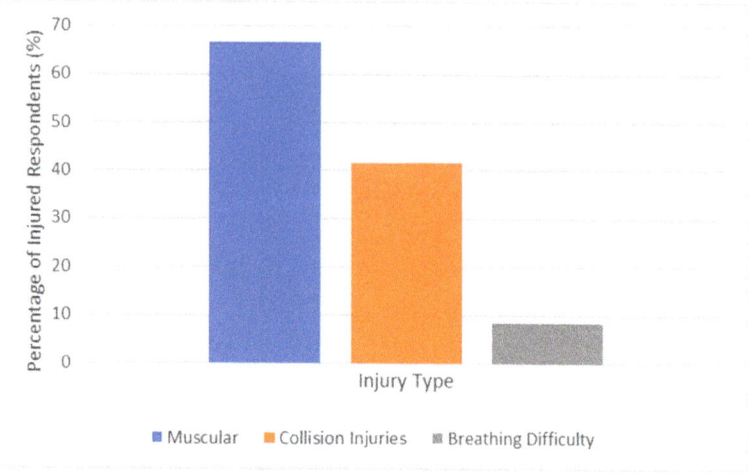

Figure 2. Driver injury types—percentage of injured respondents.

2.2. Passenger Interviews

2.2.1. Passenger Profile

Firstly, it was important to identify the types of rickshaw users to recognize usage patterns. In total, 60% of people asked used the service daily. It was found that the typical journey was approximately 1.5 km and lasted a duration of 11 minutes. This correlated closely with the information taken from the drivers in Section 2.1.2.

2.2.2. Passenger Opinion

Two-thirds said they like the convenience and availability of the rickshaw, while 20% liked the heritage and tradition associated with the rickshaw, and 13% preferred the service over walking. Negatively, 40% raised safety concerns from design to vehicle condition. Poor comfort was a reason of dislike for 20% of the people asked, while 20% found the service was too hard on the drivers, and they, in turn, felt ashamed while using the service. The attitude and behaviour of drivers was a concern for 20% or the interviewees, while a lack of luggage space was a problem for 13% (see Figure 3). When asked about their overall satisfaction of the service, it was averaged to be 7.3/10.

Although the level of satisfaction could be deemed as relatively high, the question of what improvements passengers would like to see implemented into the current rickshaw design was put forward. It was found that 60% of people desired better comfort in terms of seating and shade, while 47% thought better standards of safety, namely seat belts, would make an improvement. Further, 20% thought the idea of making rickshaws motorised would increase their satisfaction of the service, and 13% required more luggage space.

The question was put whether access to a USB charging point would be useful in their journeys and would they be willing to pay more for it: 53% claimed it would be a good idea while the rest suggested it would be futile due to too short a journey duration. Of the 53% that thought it would be beneficial, 75% would be willing to pay more for the service.

When asked if they would pay more for the use of an E-rickshaw over the conventional cycle rickshaw, 80% said they would although 25% of them said only if there was an improvement in service.

Combining the data collected from both sets of interviews, the needs and wants of the rickshaw community can be assessed and considered for design. Without this new first-hand data, the design would have solely relied upon and related to older and less focused data.

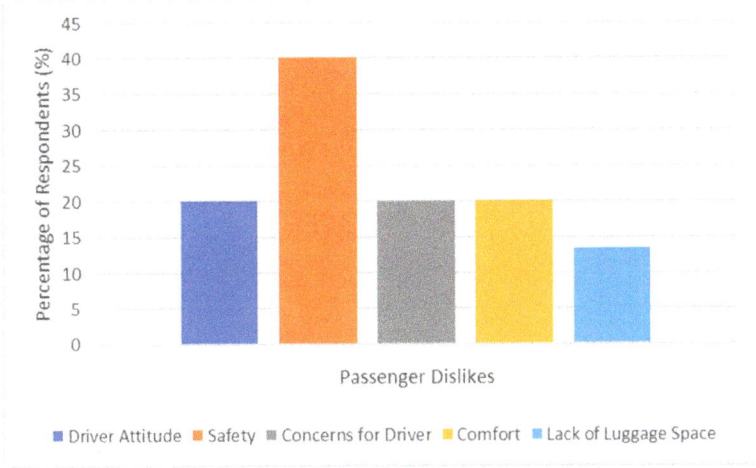

Figure 3. Passenger dislikes.

2.3. Summary of Data

In summary, it was found that the drivers work long hours, almost every day, for an exceptionally low wage, and they are seeking design improvements to lessen the physical burden to help in the prevention of injury. The passengers, although generally satisfied with the service provided, are typically seeking design improvements to alleviate safety concerns, increase comfort and protection from the weather, and to become more practical in terms of luggage space.

3. Performance Testing of the Rickshaw

An important aspect of the research conducted in Kolkata was the testing of current rickshaws by collecting performance and topography data from journeys onboard a standard cycle rickshaw.

In terms of performance, it was important to identify the current typical standard of a rickshaw, mainly maximum speed and average speed, in order to produce a vehicle capable of at least matching these parameters. It is not necessarily important to better these figures as the vehicles are typically found in crowded areas with many pedestrians and other vehicles. Therefore, the lack of space and concerns about safety do not allow the vehicle to be driven at high speed. Methodology of testing can be found in Appendix B.

3.1. Speed Testing Results

The highest maximum speed recorded was 13.54 km/h while the highest average speed was 7.77 km/h. Testing was carried out in different locations, at different times of day, with different drivers, to allow varied parameters, giving the scope to produce the most diverse data, and therefore a higher likelihood of finding the typical performance standards of the current vehicle. It is important that the design is capable of meeting the needs of every journey. Therefore, a figure of 15 km/h maximum speed and an average speed 8.5 km/h will be assumed. This provides a margin of error in the case that the results were not exactly representative of typical journeys.

3.2. Topography Testing Results

The maximum gradient was found to be 2.06%. This was in the Rawdon Street Area, the journey mapped out as shown in Figure 4. Identifying the top speed and average speed allows the research to get an overview of the basic performance parameters that the vehicle must at least equal.

Figure 4. Journey in Rawdon Street.

3.3. Energy Usage

Further to the data collected, two full journeys were mapped in terms of speed throughout; this allowed the calculation of energy used, from which the typical km per Ah and full expected range could be calculated, using a simulation software produced in Microsoft Excel by Professor T Muneer, of Edinburgh Napier University [8]. It should be noted the term SCx in Table 1 is representative of drag area, the product of frontal area, and drag coefficient.

Table 1. Simulation data.

Vehicle Characteristics		
Component	Value	Unit
Dimensions		
overall length	2.3	m
width	1.325	m
height	1.65	m
Frontal area	2.186	m^2
SCx	1.84	m^2
Drag Coefficient	0.842	
Load	150	kg
Standard person weight	70	kg

3.4. Results

Figure 5 shows the speed/time graph for area 1—Park Circus, from which it can be seen the top speed was 13 km/h. The simulator calculated that the energy used based on the conditions set in Tables 1 and 2, as 0.052 kWh. Knowing the kWh used in the 1 km journey, an estimation of the full 42 km expected range can be calculated as 2.142 kWh. The average wattage can be calculated as 304.5 W.

Table 2. Simulation situation inputs.

Simulation Inputs		
Component	Value	Unit
Total mass (with 3 people)	360	kg
Rolling friction factor	0.014	
Density of air	1.1644	kg/m^3

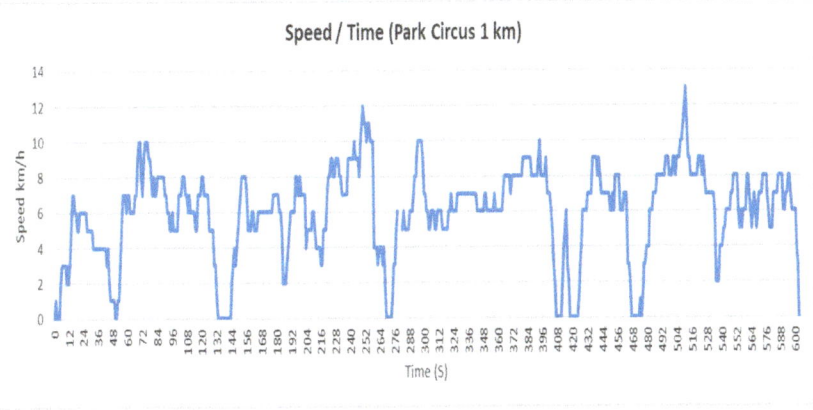

Figure 5. Park Circus speed/time graph.

Figure 6 shows the speed/time graph for area 2—Behala, in which the same testing is shown. The results of the testing showed the top speed was 13 km/h. The energy used was 0.06 kWh; therefore, for the 42 km range, a total of 2.52 kWh, with an average wattage of 385.7 W.

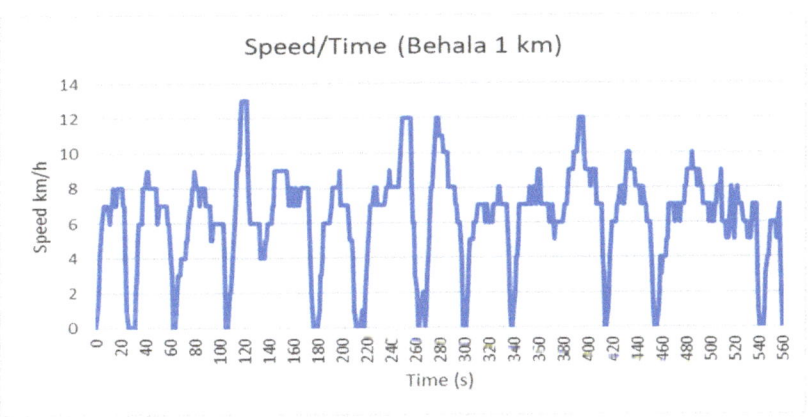

Figure 6. Behala speed/time graph.

4. Design Brief

To implement an electric motor onto the current cycle rickshaw design and improve upon current design in terms of safety and performance, the design must be environmentally friendly, built using the drivers' current rickshaw as a base, and sourcing recycled parts where practical. It must meet the daily demands of the rickshaws drivers and reduce the risk of injury by reducing the physical efforts needed to operate it. It must meet the standards of what the customers require in a rickshaw. It must remain within the boundaries of the traditional design aesthetically. A detailed product design specification (PDS) can be found in Appendix C.

5. Design

5.1. Solar

As the vehicle is powered though the use of batteries, they must be charged; since the vehicle is to remain environmentally friendly, the charge must be supplied by a renewable energy source. The source identified as a potential is solar power.

5.1.1. Solar Availability

The average global horizontal irradiance of Kolkata is as shown in Figure 7, giving the researcher an indication of the highest and lowest expected solar availability by month. This allows calculation to be carried out to predict if the on-board solar panel will fully charge the vehicle each day [9].

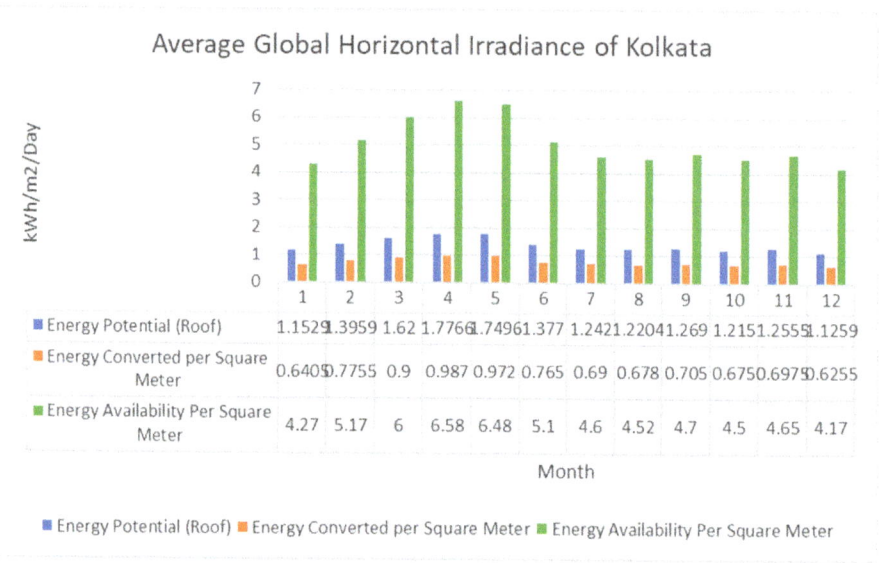

Figure 7. Average global horizontal irradiance of Kolkata.

5.1.2. Solar Farm

An option would to be have a solar farm outside of the of the built-up area. This would be of large scale and would be capable of supplying the whole community of E-rickshaw drivers with fully recharged batteries every day. The drivers would exchange their flat batteries with the fully charged batteries. This would enable them to power their rickshaw each day. In order to facilitate this, it would have to be fully funded by government or a private company in return for a fee from the drivers. To seek this option, funding must be acquired, but it remains the only feasible method of recharging using solar power as outlined.

5.2. Funding

Other possible ways of funding this could include sponsorship where companies could bid to sponsor a number of rickshaws. In return, the company would be able to place their advertisement on each rickshaw. This could either help fund the solar farm or could help fund cost to the drivers in terms of replacement battery fees. Although speculative, this would be a possible revenue stream that should be explored.

5.3. Electrical Components

5.3.1. Drive Systems

When implementing an electric drive onto a bicycle there are two options, either fully electric powered or pedal assisted.

Fully electric powered (FE) with all power coming from the motor: This method requires no physical human effort, except speed control, using a throttle.

Pedal-assisted power (PA) is where the motor and driver both work together and share the burden of the work.

Each drive system has its advantages and disadvantages when compared to each other. Fully electric powered is the least demanding upon the driver but has a significantly reduced range, as it takes the full load at all times; therefore, to compensate this, an additional battery capacity would be needed at an increased cost. Pedal assist gives the driver electric drive assistance, lowering the physical demand without completely removing it; this keeps an aspect of exercise, therefore retaining associated health benefits.

As seen in Table 3, a drive selection matrix shows the design matrix comparison between each drive type, and through analysis, a pedal-assisted drive system will be used for the design, as it scores higher in the key parameters.

Table 3. Drive selection matrix.

Parameter	Weighting	Drive Type	
		FE	PA
Cost	2	4	8
Performance	1.5	9	7
Ease of Installation	1	8	8
Maintenance Required	1.5	7	7
Reliability	2	8	9
Safety	2	9	9
Range	2	5	8
Total		77	90

5.3.2. Assist Sensor

A sensor must be fitted in order to provide the required input of assist from the motor; there are different options to what type of sensor can be used. Most commonly, on pedal-assisted e-bikes, there are two options, a cadence sensor or a torque sensor. Each sensor type has its own merits and limitations; therefore, each must be assessed to find the most suitable component for the rickshaw application [10].

The implementation of a cadence sensor will allow the controller to detect the Revolutions Per Minute (RPM) of the pedal crank, thereby being able to send a signal to the motor to turn on. Assistance levels can be modified through the use of a digital display control computer. This allows the driver to choose a setting to match the requirements of the road condition, or for battery conservation. Cadence sensors require the driver to have pedalled approximately one-half rotation of the crank before the motor is able to detect the human input; therefore, the power is slightly delayed. Cadence sensors are inexpensive, easily installed, and require little maintenance.

5.3.3. Control Unit

A control unit is needed to manage each electrical component; the unit will receive signals from any activated sensor and it will then send the required output signal to whatever component is related to the activated sensor. The control unit will control the speed of the motor, based on the cadence sensor, by restricting/enabling the power provided by the batteries; it will also provide power to any lighting when the drive engages the button or switch to turn them on.

5.3.4. Motor

As the results of testing in Section 3.3 showed a highest average of 385.7 W, the motor must be capable of achieving this output. The results were of average wattage; therefore, a larger motor will be needed to meet the peak power demands. The testing was not sufficient to accurately identify peak power as energy expelled, though acceleration up gradient could not be calculated accurately as the increments of gradient were too large; therefore, an estimated simulation was carried out to show a reasonable indication of the peak power.

A simulation was executed where it was assumed the driver was accelerating up a slope of 2.06% as per the maximum discovered in testing, assuming they would reach a top speed of 13 km/h in 10 s, though testing this was never achieved, thus being an exaggerated scenario. This will give an overestimation of peak power; therefore, the wattage should be an acceptable measure for a motor that can replicate all human power.

The result of the simulation showed a wattage of 468 W; therefore, a 500 W 24 V DC motor is an acceptable choice for the design.

5.3.5. Battery

Arguably, the battery is the most important component of an electric vehicle, it has a huge bearing on the range capabilities. There are many considerations that must be satisfied when choosing a battery option such as, cost, ampere hours, voltage, depth of discharge, cycle life, physical size, and weight.

From testing in Section 3.3 it was discovered that the motor will require a total of between 2.142 kWh and 2.52 kWh. Taking the higher usage, the Ah capacity can be calculated as 105 Ah.

As stated in the PDS (see Appendix C), the main considerations in the design of the e-rickshaw are cost and range; therefore, the chosen battery must reflect that. Firstly, as cost is the major limitation in this design, the cost of a lithium battery system suitable for the design could cost as much as £1500, based on typical retail prices (March 2019). This consequently rules out the possibility of using lithium batteries in the design.

A suitable lead acid battery system would cost around £200 based on typical retail prices (March 2019), which consumes the majority of the budget; therefore, the sourcing of inexpensive refurbished or second-hand units may be the most viable option.

Due to the depth of discharge in lead acid batteries being around 50%, achieving a life cycle of 1000 cycles [11], a system of 4110 Ah units will be installed in a circuit of 2 parallel branches; this will provide the motor with the 24 V needed and a capacity of 210 Ah, assuming a depth of discharge (DOD) of 50% a usable capacity of 110 Ah, which exceeds the targeted range.

5.4. Mechanical Components

5.4.1. Gearing

Commonly in India, the rickshaws are fitted with a singular gear and therefore unable to change torque for different gradients and speeds; this is largely due to cost reduction, lower maintenance, and for simplicity of design. This fits with the PDS brief and therefore should be continued. The pedal sprocket transfers the human power onto the rear shaft via a link chain and a gear attached to the shaft, and the power can be stepped up or stepped down by a multitude of gear sizes. Typically, on a rickshaw the ratio is 1:0.56 and the pedal sprocket of 48 teeth is stepped down to a 27 tooth on the rear shaft [12]; therefore, for every one full pedal cycle, the rear shaft rotates 1.778 times; the rickshaw, neglecting any losses in grips at the wheels, should move 1.778 times the circumference of the rear wheels.

With this ratio, a pedal sprocket cadence of 90 RPM would propel the rickshaw at a speed of 20 km/h, and this is acceptable for the final design; although, this ratio does not suit acceleration, and therefore must be replaced with a ratio that reduces the physical efforts of the driver. A ratio of 1:1 will provide the driver with increased torque allowing the easier rotation of the pedals and will drive at a speed of 11 km/h at a cadence of 90 RPM. It is important to keep the gearing ratio to a reasonably achievable speed; if an electric component fails, then the driver may still be able to cycle manually and either complete his shift or take the rickshaw to a repair centre.

The gearing of the motor will be set to provide the vehicle with a top speed of 15 km/h; since the rated RPM of the motor is 2500, the gearing must step the RPM down to 192, using a gear ratio 1:13 this. In theory, can be achieved.

5.4.2. Free Wheels

The gears on the shaft must be able to work independently of each other; this is done by installing freewheel sprockets. A freewheel sprocket is a type of overrunning clutch, which allows the shaft to turn faster than the gear, meaning the pedals can be stationary while the shaft rotates. If the gears were fixed, the pedals would consequently be powered by the motor, causing a high risk of injury to the driver. Installing a freewheel on each gear will also allow the driver to pedal faster than the motor, in case of low power input or electrical failure.

5.4.3. Wheel Size and Tyre Width

The coefficient of rolling resistance can be reduced by the wheel and tyre choice; as a general rule, the larger the wheel diameter and wider the tyre thickness, the lower the coefficient of rolling resistance will be [13]. In the case of the design, the donor rickshaw wheels will be analysed for dimensions as a slight increase in the coefficient of rolling resistance can increase the energy used considerably. The condition will also be analysed as warped or damaged wheels will also contribute to wasted energy. A comparison of a coefficient of rolling resistances 0.015 and 0.012 for the area of Park Circus was drawn, with total of 42 km (the expected daily range). It was calculated that with a coefficient of rolling resistance of 0.014, it uses 0.004 kWh/km more, 0.168 kWh over 42 km, thus the battery capacity will be depleted by an extra 7 Ah; therefore, it is crucial to have the most efficient wheel and tyre choice available.

5.5. Safety

As highlighted by the research conducted in Kolkata, there were many safety concerns raised by both the drivers and passengers alike. Concerns such as potential injury for impact with other vehicles, uncomfortable seating causing soreness and bruising, and insecure seating allowing the passenger to move around in the seat unwillingly, potentially causing harm. It was also said that the rickshaws can be awkward to embark due to no handle to grip, causing the potential of falling or slipping. Safety concerns need to be addressed.

5.5.1. Seating

Firstly, in relation to the insecure seating, many passengers suggested seat belts; this would be a simple application for a conventional vehicle. However, the rickshaw is used quite differently. In a standard car, the seating is individual and is provided with supplementary seat belts, but this is not the case for the typical rickshaw that incorporates a bench-style seat. This proves to create more complication in design as the seating is designed for one or two people, but in reality, it has to function for multiple purposes, as occasionally drivers will take more than two if there are small children.

This led to the idea of modifying an automotive style seat belt to be capable of adjusting to 1 or 2 or more people. The seat belt will be able to be plugged into two different locations, allowing the adaption to fit one or more passengers. It will also be adjusted as per standard automotive seat belts to securely and comfortably restrict movement of the passenger. The seat belts could be sourced from automotive dismantlers to ensure a lower cost than a new unit.

The seating as a unit can be improved simply by adding thicker or softer padding to the base back and sides. Many of the rickshaws tested, while in Kolkata, were fitted with insufficient or worn-out padding. This led to the ride comfort for the passengers being limited; this was stressed by 20% of the passengers interviewed and also commented on by drivers who were aware of the issue.

5.5.2. Handle

The addition of a handle can be easily incorporated into the curb-side of the frame; this is a simple and inexpensive, yet practical solution for the aid of embarking the rickshaw. The handle will be mild steel tubing bent to shape and welded directly onto the frame.

5.5.3. Collision Prevention

The concern of risk of injury during impact will already be significantly reduced for the passengers with the addition of more padding and a seat belt, but the risk to the drivers still remains. This is a concern that is exemplified by the total of 42% of the injuries reported amongst the drivers interviewed. In an ideal scenario, a protective shell would be constructed to protect the occupants from collisions. Unfortunately, due to financial restrictions, the fix may not be the cure but the prevention. Currently, the drivers signal their intent of direction with their arms; not only is this a danger as other vehicles passing could collide with the driver's arm causing injury, it also prevents the driver being in full control of the rickshaw, by impeding their ability to steer correctly. The addition of indicators would allow the driver to keep both hands on the handlebars at all time while being able to alert the other road users of their intended direction of travel. The indicators will be comprised of a small cluster of LEDs; this will make their battery usage negligible.

Although this solution will not have any effect on injury prevention when there has been a collision, in theory it should lower the risk of a collision taking place.

In addition to the implementation of indicators, reflective strips will improve visibility of the presence of the rickshaw in low light, thus reducing the risk of collision further [14].

5.5.4. Shade Design

The climate of Kolkata is that of extreme high temperatures and heavy rainfall. Data from 1901–2000 shows that the average monthly rainfall for July can reach 339.8 mm [7]. Heat stress can cause many unwanted bodily reactions ranging from sweating to the failure of control mechanisms [15]; as this can become a serious threat to health, shading must provide protection from the heat.

The UV radiation over India is exceptionally high. The UV index can reach as high as 16. This is exceptionally hazardous to the health of anyone exposed to this for any prolonged length of time [7]. Ultraviolet radiation is damaging to the skin and over exposure can lead to skin cancers. This causes a major risk to rickshaw drivers as they work an average of 11.36 h per day. Currently, the driver has no shade and the design only caters for the passengers; therefore, a shade must be installed that caters for all occupants. The shade must improve defence against UV rays, form shelter from the heat and heavy rainfall, but it must not add excessive weight, cost, or drag.

Table 4 shows the design matrix of the range of materials explored.

Tarpaulin will be used in the design as it provides the best performance in the key areas scoring better than aluminium, which although performs better in many areas, it simply costs too much to be used.

As each donor rickshaw will have different frame designs, a standard roof design will be fabricated to fit through the use of tubing; an example of an outcome of the shade is shown in Figure 8.

Table 4. Shade material selection matrix.

Parameter	Weighting	Drive Type		
		Aluminium	Cotton	Tarpaulin
Cost	2	1	9	8
Water Proofing	2	10	1	9
Ease of Installation	1	8	9	9
Maintenance Required	1.5	10	4	7
Aesthetic Value	1	9	7	5
UV Protection	2	10	9	10
Heat Barrier	2	4	5	7
	Total	74	63	84.5

Figure 8. Example of roof panel fabrication.

This would conform to all the parameters set as the fabric is very low cost and the shade fully covers all occupants protecting them from extreme weather conditions. The rear of the shade would feature a zip or series of fasteners to allow a side and rear section to be attached. This gives protection from rain whilst up, and reduces the drag whilst down, allowing the driver to choose and employ a setup quickly to cater for the needs of the conditions.

5.6. Servicing

5.6.1. Battery Compartment Design

As the batteries must be removed to be charged and replaced, it is vital that they are easily accessible; therefore, strategic location of the battery units must be considered. There is a large unused space under the passenger seating and this would be an ideal location to house the batteries as it can be easily closed off for safety and it is also located close to the motor and driven shaft, therefore the cabling will also be enclosed and will be shorter, consequently reducing costs. The batteries would be accessible from a hinged door at the rear of the vehicle, which will be laid out to maintain an even weight balance. This will ensure the weight is supported evenly throughout the vehicle to ensure that components are not subject to uneven wear. It will also maintain handling of the vehicle through cornering.

The batteries will be secured to the base of the rickshaw by a bracket (as seen in Figure 9) which will be connected by a threaded rod and nut to the base of the rickshaw battery compartment. This will allow quick release of each battery allowing a more efficient changeover of batteries, it will also prohibit movement of the batteries while in operation.

Figure 9. Bracket to secure battery.

5.6.2. Storage

A luggage area was requested by 13% of the passengers asked. As space inside the cabin of the rickshaw is limited, a rear storage box will be installed on the rear exterior frame of the cabin. The storage box will be fabricated from a mild steel frame welded to the rickshaw frame and will be enclosed with plywood with a hinged door for access. A standard storage compartment will be fabricated to fit each individual rickshaw

5.7. Expected Cost

Table 5 shows an indication of expected pricing and target cost for each component, totalling £235, but this would have to be verified through contacting wholesalers and manufacturing factories to obtain exact costing. Labour costs must be considered, therefore using the average wage of 239,465 rupees (£2750) for a mechanic in India, the hourly rate based on a 40-h working week can be considered as being approximately £1.32. The fabrication and modifications to the rickshaw are basic and should require no longer than 25 h; therefore, there is an expected labour cost of £33. Machinery such as welding equipment and lathes will be required; as these will be used to produce thousands of rickshaws, the cost per unit will be negligible and therefore overlooked at this stage of costing. A reasonable approximation of cost would total £268.

Table 5. Design specification and costing.

Component	Specification	Expected Cost
Motor	500W 24 V DC	£30
4 × Lead Acid Battery	110 AH	£100 (used)
Cadence Sensor	N/A	£3
2 × Freewheel Sprockets	1 × 27 T 1 × 170 T	£20
2 × Gear Sprocket	1 × 27 T 1 × 10 T	£20
Control Unit	24 V	£9
Seat Belt	N/A	£5 (used)
LED Lighting	2.25 V	£2
Cabling	Various amps	£1
Shade Construction	Steel/Tarpaulin	£30
Storage Construction	Steel/Plywood	£15
Total		£235

6. Proof of Concept Prototype

A prototype was developed to have a working model of the drive system. The prototype was not built to resemble a rickshaw or to carry the same weight as what would be expected, it was purely to test the electrical and mechanical components and their configuration within the design.

6.1. Development and Fabrication

The prototype was developed using the frame of a mountain bike; this would provide the base to be built upon. The donor bike had a wheel diameter of 24 inches.

The bike was stripped down to the frame, front wheel, front brake, and pedal and pedal gears, of which the 24-tooth gear would be used. As the rear wheel would be replaced by two wheels, a shaft had to be designed to facilitate this; a 19 mm mild steel metal rod was fabricated to suit.

It was vital the shaft could house bearings to separate the shaft from the frame; a length of stainless steel of a 32 mm outside diameter and inside diameter of 28 mm was chosen and this allowed the selection of bearings. Four bearings of 17 mm inside diameter and 28 outside diameter were purchased; this required the shaft to be lathed to enable the bearing to have a press fit onto the shaft as seen in Figure 10. Gearing was to be attached within the central portion of the shaft. The precise measurements were mapped out and the bearing positions could be identified.

Figure 10. Bearing press fitted onto shaft.

The gearing was set for a ratio of 1:0.75 for the human input and 1:1.29 for the motor. This would give the driver a relatively undemanding start from 0 km/h and would provide a speed or 15 km/h at a cadence of 90 rpm (unassisted), which is regarded as ideal for prolonged cycling [16].

The 18 tooth gears bought came in the form of a freewheel assembly. This allows each drive method to work independently providing safety to the driver.

Having assembled the shaft (see Figure 11), a frame could be developed to attach the shaft to the front portion of the bike; using mild steel tubing of 26 mm outside diameter and 20 m inside diameter, the frame was metal inert gas (MIG) welded to the axle, and a platform section of frame was developed (see Figure 12) in order to locate the motor, batteries, and control unit.

Having the platform section of frame, an 18 mm sheet of plywood could be cut to size and be attached to provide a base for the electrical components (see Figure 13).

Figure 11. Rear shaft and axle assembly.

Figure 12. Platform frame assembly.

Figure 13. Platform housing electrical components.

The electrical components to be installed were 1 × 24 V DC motor, 2 × 12 V 45 AH lead acid batteries (used but fully charged), 1 × 24v control unit, one cadence sensor, and one three-phase bridge rectifier. A slot directly above the motor driven gear was cut providing a space facilitating the link chain from gear to motor (see Figure 14); the motor was then secured in place, and the use of spacers allowed the tension of the chain to be set.

Figure 14. Slot for chain to motor.

The cadence sensor was fitted on the non-geared side of the pedal sprocket (see Figure 15), the cable was cable tied for safety, and ran back to the platform. The control unit was secured by cable ties to the plywood and the batteries were secured using screws. The batteries were connected in series and to the control unit to provide the 24 V needed to run the motor; the cadence sensor was then connected. Finally, to connect the motor to the control unit, a three-phase bridge rectifier was used to convert the three-phase signal from the control unit to the single-phase DC motor.

Figure 15. Cadence sensor assembly.

The finished prototype is shown in Figure 16.

Figure 16. Finished prototype.

6.2. Testing

Placing the prototype's rear axle on stands took all weight off the wheels and let the tester record the activation of the motor and to prove the concept. The pedal sprocket was turned by hand while the motor and pedal position were monitored visually by camera recording; this was tested in the same styles as the road testing where low, moderate, and high cadences were tested three times each. It was found that the motor would activate after one half rotation of the pedal; this, in theory, would give the driver assistance after a total of 0.42 m, therefore alleviating all effort from then through acceleration to desired speed.

Through calculation it was discovered that the motor would benefit from a higher gear ratio. Currently, the prototype has a ratio of 1:1.29; a ratio of 1:14 will give the prototype a theoretical top speed of 15 km/h, which is the desired minimum top speed set in in the PDS see Appendix C.

6.3. Limitations

The testing of the motor activation was not accurate as no angular measurement was taken and was visually inspected only. The testing was for indicative purposes to prove the concept. The vehicle did not replicate a rickshaw exactly, therefore although this method of testing is acceptable at this stage, more-developed prototypes will be more accurately tested in the future.

7. Evaluation

7.1. Kolkata Research and Testing

The results of the research and testing in Kolkata were successful to a degree; the research proved invaluable as it was a direct first-hand source that gave the project a focused viewpoint directly identifying issues to be developed in design; the testing of the rickshaws proved to provide semi-accurate results and the inaccuracies came from the poor gradient analysis of the application used. This would have been improved by using a more precise tool for measuring altitudes. If funding allows it, this would be used in future testing.

7.2. Design

The design of the E-rickshaw, although not ground-breaking in terms of technology, meets the demand of the PDS. Therefore, the design can be classed as a success, and can be further advanced in development in order to get to a working prototype stage.

7.3. Proof of Concept Prototype

The gearing was also configured insufficiently and this resulted in inconclusive road testing. The concept was still proven as the electrical components responded as expected, therefore setting a foundation for further development. Therefore, as a proof of concept, it was a relatively successful tool for analysis.

8. Conclusions

As a whole, the project was a success and proved that improvements to the design could make a rickshaw less labour-intensive to drive, safer, and more comfortable without compromising on the environmentally friendly quality that cycle rickshaws possess, while remaining relatively inexpensive. Although the rickshaw would be inexpensive to construct, the feasibly of the solar aspect has not been explored and must be carried out before the project can be classed as a total success. If the solar aspect is not feasible, the project would not meet the criteria set in the PDS as a solar-sourced electric vehicle, although the vehicle could be charged by other means.

9. Recommendations

As it has been concluded that the project has potential to be a success, it would be recommended that further development of a working rickshaw could be prototyped and tested, ideally in location. This would get the project to a point where funding could be sought, as funding is the key to the objectives of this project being able to become applied.

Author Contributions: Conceptualization, T.M. and C.C.; methodology, T.M. and C.C.; software, T.M.; validation, T.M. and C.C.; formal analysis, C.C.; investigation, C.C.; resources, C.C. and B.F.; data curation, C.C.; writing—original draft preparation, C.C.; writing—review and editing, C.C. and T.M.; visualization, C.C.; supervision, T.M.; project administration, C.C., T.M. and B.F.; funding acquisition, C.C. and B.F.

Funding: This research received no external funding.

Conflicts of Interest: The authors declare no conflict of interest.

Appendix A

Interviews:

- Interview methodology

The conduction of the interviews was assisted by The Neotia University, Kolkata; as the researcher was not able to communicate in the local language, Bengali, students acted as interpreters. The students asked the drivers the set questions and relayed the answers to the researcher in English. The interviews took place in various localities of the city.

- Driver Profile

Questions relating to social aspects were configured to determine a brief profile of the drivers of Kolkata, India. It was decided that age must be acquired as an indication of premature retirement due to injury; combining age with the questions of how many hours and days worked, a comparison between the two could be established to investigate if there was a correlation between age and work pattern. Knowing the working hours of the drivers, the question of expected daily income could give an insight into the financial position of the drivers, and it could be used to evaluate their ability to fund improvements to their rickshaws.

- Passenger Profile

Local residents who used rickshaw services were interviewed in order to identify the usage patterns, overall satisfaction, needs, and wants from a customer's point of view. A total of 15 people were interviewed, with each being asked eight questions.

- Limitations

The limitations of the interview process include: The small quantity of interviewees asked, possibly exaggerating abnormalities; relying on the honesty of the interviewees especially where the questions were relatively personal and intrusive; the language barrier may have given the opportunity for miscommunication and misunderstanding to arise; the questions could contain confirmation bias leading unconsciously to answers the researcher was hoping for.

Appendix B

Testing

- Speed Testing Methodology

Data from journeys at different locations around the city were taken in order to ensure a more balanced scope of topography and speeds was achieved. The drivers were chosen at random and were of differing ages and physiques; this gives the results the greatest chance for variation in order to find as true an average as possible. The areas where journeys were tested were Ripon Street, Behala, Park Circus, South City, and College Street. The data were collected using the mobile GPRS application 'Elevation tracker' recorded on an iPhone 6. The application provided: Distance travelled, maximum gradient, average gradient, maximum speed, average speed, and also mapped out the journey.

- Topography Testing Methodology

The topography was also mapped using the same mobile application, 'Elevation Tracker'. It is vital to the power requirement calculation that the highest max gradient is identified. In testing, it was impossible to test the topography of all roads in Kolkata used by the rickshaws due to time and monetary constraints, but varied journeys in various locations gave a good representative sample.

- Energy Usage Methodology

In order to collect the data, two journeys of 1 km were undertaken on two separate rickshaws with two different drivers in two different locations (Park Circus and Behala). This gave the opportunity for the data to show a variation of results. Each rickshaw driver was chosen at random and asked to drive 1 km, and this was measured using the mobile app, 'Gradient Tracker'. The speed was recorded using another mobile application 'Speedometer' on a sperate device; using the application 'Screen Recording', a video of the speeds would be created for the entirety of the journey. Using Windows Media Player, the video was paused and skipped in one second increments; at each second of the video, the speed shown was inputted into the Excel simulation. The simulation programme provided the average speed, distance travelled, duration, and energy used; further to this, it was used to create a speed/time graph to show the acceleration of the rickshaw. The simulator was set to replicate the conditions of the journey by imputing the relevant values as shown in Tables 1 and 2.

Appendix C

Product Design Specification (PDS)

Performance:

- It must be able to reach a speed of at least 15 km/h.
- It must have a range of at least 42 km
- It must be able to carry 2 passengers and a driver

Aesthetics

- The rickshaw should be aesthetically pleasing and fit within the heritage.

Ergonomics:

- It must be comfortable to pedal.
- It must be designed to reduce risk of injury.
- All moving parts must be contained safely.
- It must be easy to embark and disembark for the passengers.

Environment:

- The rickshaw must protect the occupants from the weather where necessary.
- It must be able to operate efficiently in the climate.
- It must be charged using solar power.

Quantity and Manufacture

- It must be able to be manufactured in a mass quantity inexpensively.
- It must be able to be manufactured without the need for specialised machinery or tooling requiring high level workers.

Market constraints:

- It must merit the higher cost and price to compete with the cycle rickshaw market.

Competition:

- It must be able to be priced significantly cheaper than the £1100 Toto E-rickshaw [17] to avoid competition with this particular comparable vehicle.

Quality and consistency:

- It must be reliable and durable.
- It must require low maintenance
- Each vehicle should have identical performance when built.

Materials:

- All materials used must be capable of withstanding corrosion from the climate.
- The materials should be able to last the expected lifespan of the rickshaw or be able to be replaced inexpensively.
- The rickshaw should last over 20 years with maintenance and component replacements.

Vehicle constraints

- The total vehicle weight must not exceed 150 kg.

Battery requirements

- Must have a recharge time of less than 24 h.
- Must not exceed total weight of 40 kg.
- Must have be a usable 110 Ah of charge.
- Batteries must be easily accessible and removable.

Standards:

- It must adhere to all rules and regulations for the location at which it is situated.

Target product cost

- The product must cost less than £350.

Patents

- No patents relating to the E-rickshaw have been found in the Indian directory.

References

1. Chandran, N.; Brahmachari, S.K. Technology, knowledge and markets: connecting the dots—Electric rickshaw in India as a case study. *J. Frugal Innov.* **2015**, *1*, 2–10. [CrossRef]
2. Khan, J.H.; Hassan, T. Socio-economic profile of cycle rickshaw. *Eur. Sci. J.* **2010**, *8*, 320.
3. Registrar General & Census Commissioner. *Census of India 2011*; Registrar General & Census Commissioner: New Delhi, India, 2011.
4. Goswami, D. *Understanding the Socio-Economic Condition of Rickshaw Pullers: After Arrival of E-Rickshaw*; University of Delhi: New Delhi, India, 2017.
5. Ranjan, C.; Akhtar, S.J.; Tiwari, A.K.; Choubey, N.K.; Kumar, R.; Rajwar, L.; Kumar, N.; Kumar, R. Desing and fabrication of tri rickshaw with solar charging station. *Int. J. Mech. Eng. Technol.* **2017**, *8*, 358–366.
6. Attri, S.D.; Tyagi, A. *Climate Profile of India*; India Meteorological Department Ministry of Earth Sciences: New Delhi, India, 2010.
7. Bhattacharya, R.; Pal, S.; Bhoumick, A.; Barman, P. Annual variability and distribution of ultraviolet inder over India using temis data. *Int. J. Eng. Sci. Technol.* **2012**, *4*, 77–83.
8. Muneer, T.; Kolhe, M.L.; Doyle, A. *Electric Vehicles Prospects and Challenges*, 1st ed.; Elsevier: Edinburgh, Scotland, 2017.
9. Solar Energy Centre, MNRE, Indian Metrological Department. *Typical Climatic Data for Selected Radiation Stations*; Solar Energy Centre: New Dehli, India, 2008.
10. Van Ditten, M. *Torque Sensing for E-Bike Applications*; Delft University of Technology: Delft, The Netherlands, 2011.
11. Albright, G.; Edie, J.; Al-Halla, S. *Comparison of Lead Acid to Lithium-ion in Stationary Storage Applications*; All Cell Technologies LLC: Chicago, IL, USA, 2012.
12. Hickman, M.R. *A Study on Power Assists for Bicycle Rickshaws in India, including Fabrication of Test Apparatus*; Massachusetts Institute of Technology: Cambridge, MA, USA, 2011.
13. Steyn, W.; Warnich, J. Comparison of tyre rolling resistance for different mountain bike tyre diameters and surface conditions. *S. Afr. J. Res. Sport Phys. Educ. Recreat.* **2014**, *36*, 179–193.
14. European Cyclists Federation. *Requirements on Lighting (Light Intensity) and Reflectors of Bicycles*; ANEC: Brussels, Belgium, 2012.
15. Health and Safety Executive. *Heat Stress in the Workplace*; Health and Safety Executive: London, UK, 2013.
16. Abbiss, C.; Peiffer, J.; Laursen, P. Optimal cadence selection during cycling. *Int. Sportmed. J.* **2009**, *10*, 1–15.
17. IndiaMart. 2019. toto-e-rickshaw-16669287130. Available online: https://www.indiamart.com/proddetail/toto-e-rickshaw-16669287130.html (accessed on 2 February 2019).

© 2019 by the authors. Licensee MDPI, Basel, Switzerland. This article is an open access article distributed under the terms and conditions of the Creative Commons Attribution (CC BY) license (http://creativecommons.org/licenses/by/4.0/).

Article

Driven Performance of Electric Vehicles in Edinburgh and Its Environs

Ross Milligan [1], Saioa Etxebarria [2], Tariq Muneer [3] and Eulalia Jadraque Gago [4,*]

1 Edinburgh College, Edinburgh EH11 4DE, UK
2 Department of Mechanical Engineering, University of the Basque Country, 01006 Vitoria-Gasteiz, Spain
3 School of Engineering and the Built Environment, Edinburgh Napier University, Edinburgh EH10 5DT, UK
4 School of Civil Engineering, University of Granada, 18071 Granada, Spain
* Correspondence: ejadraque@ugr.es

Received: 3 July 2019; Accepted: 5 August 2019; Published: 9 August 2019

Abstract: Fuelled by energy security problems and urban air pollution challenges, several countries worldwide have set the objective to gradually eliminate petrol and diesel cars. The increasing support from government and demands for environmental friendly means of transportation are accelerating the use of battery electric vehicles. However, it is indispensable to have accurate and complete information about their behaviour in different traffic situations and road conditions. For the experimental analysis carried out in this study, three different electric vehicles from the Edinburgh College leasing program were equipped and tracked to obtain over 50 GPS and energy consumption data for short distance journeys in the Edinburgh area and long-range tests between Edinburgh and Bristol (UK). The results showed that the vehicles' energy intensities were significantly affected by the driving cycle pattern, with a noticeable diminution due to low temperatures. It was found that the real available range of the electric vehicle in some situations could be 17% lower than the predicted mileage shown in the dashboard of the vehicle. The difference from the New European Driving Cycle (NEDC) values was even higher. The study has also provided a discussion on the effect of the electricity mix on carbon emission reduction.

Keywords: electric vehicle; sustainable development; driving cycle; climate change

---

1. Introduction

Sustainable development is believed to be the solution to excessive dependence on fossil fuels, which is responsible for climate change with hazardous environmental impacts, and the potential dearth of these resources. The transportation sector, which can be categorised into subsectors including road, aviation, railway, waterways and international marine transportation, is heavily dependent on fossil fuels. The World Oil Outlook 2017 claimed that in the year 2017, this sector accounted for the 54.6% of world oil consumption [1]. According to this report, by 2040, the transport sector will consume 64.9% of the world's oil consumption, expecting an increase of 10.3% from 2017–2040.

To improve this situation, actions should focus on a modal shift to more sustainable means of transport, and on the use of more efficient and less polluting transportation through awareness raising and training.

Passenger flow forecasting is the basis for improving transport services, providing early warning for sporadic urban traffic events and making cities smarter and safer. Related to this, Liu and Chen [2] proposed a new passenger flow prediction model that uses deep learning methods to predict passenger flow per hour for Xiamen's BRT (Bus Rapid Transit) stations. Their article offered a review of passenger flow prediction models that have been developed by other authors.

Global port activity and cargo-handling operations increased rapidly in 2017, after two years of mediocre performance. According to the United Nations Conference on Trade and Development (UNCTAD), the world's container ports handled about 752.2 million- twenty-foot-equivalent-unit (TEUs) in 2017. Such a significant increase in international container trade volumes has required the marine container terminal (MCT) operators to improve terminal productivity and efficiency to meet the growing demand [3]. Dulebenets [4] proposed in 2017 a memetic algorithm with a deterministic parameter control to facilitate the programming of moorings in the MCTs, and to minimise the total cost of the ship's service. In 2018, a multi-objective mathematical model and a solution algorithm were developed that can facilitate the design of profitable vessel schedules and assist liner shipping companies with the analysis of important trade-offs between conflicting objectives.

Regarding the international maritime sector's commitment to reducing greenhouse gas emissions, the International Maritime Organisation (IMO) adopted an initial strategy to reduce annual greenhouse gas emissions from ships by at least 50% by 2050 compared to 2008 levels. For air pollution, the global limit of 0.5% for the sulphur content of fuel oil used on board ships will take effect on 1 January 2020. In order to ensure uniform application of the global sulphur limit, it is important that ship owners consider and adopt various strategies, such as the installation of flue gas scrubbers and the transition to liquefied natural gas and other low sulphur fuels [5].

In the case of air transport, during the last decade, CO_2 emissions in this industry have increased by more than 6%, with the most important source of these emissions being the fuel consumed by aircraft. Several methods have been proposed to reduce CO_2 emissions in the airline industry. Cruise control is one of the newest tools offered for this purpose, allowing aircraft speed to be adjusted [6]. Jalalian, Gholami, and Ramezanian [7] developed a non-linear multi-target mixed integer scheduling model to integrate flight scheduling, aircraft trajectory assignment and gateway assignment, with the goal of reducing CO_2 emissions and increasing the level of passenger service. These authors provided a review of previous research related to reducing fuel consumption and CO_2 emissions.

In the case of rail transport, interruptions disrupt passenger transport on a daily basis. An interruption disables the timetable, the circulation of rolling stock and the crew's timetable. Wagenaar, Kroon, and Fragkos [8], presented the concept of dead-heading travel and adjusted passenger demand in the rolling stock rescheduling problem (RSRP). The results showed that dead-heading trips is useful in reducing the number of cancelled trips.

On the other hand, the development of rail transport would help drastically reduce environmental pollution, requiring less energy consumption and less dependence on oil, thus reducing greenhouse gas emissions and reducing the use of cars and trucks. While 8% of world transport is done by rail, the volume of GHG that it generates is 3.6%.

Statistics demonstrate that the highest share of oil in the world is consumed by the road sector. According to the International Energy Agency (IEA), by 2030, oil will represent around 43% of all energy used in road transportation [9]. Consequently, road traffic greatly contributes to greenhouse gases (GHG) emissions, various health-damaging pollutants, such as nitrogen dioxide, toxicity to humans, acidification and noise nuisance [10].

It is predicted that global car sales will more than double by 2050 [11]. The annual energy demand in the European Environment Agency (EEA) member countries grew by 38% between 1990 and 2007. The energy consumption decreased 3% between 2007 and 2016. However, there was a net growth of 34% between 1990 and 2016 [12]. Energy consumption in the transportation sector has increased among all countries that are not part of the Organisation for Economic Co-operation and Development (OECD), where 80% of the global population lives. In these regions, the standard of living, and therefore the purchasing power for personal transport, rises together with the economic growth. In consequence, an average annual increase rate of 2.5% is expected in the energy consumption of the transport sector by 2040 [13].

The energetic, environmental and economic impact concerns caused by increasing global transport make it essential to work towards a more sustainable transportation system. A possible solution adopted by many national governments is the introduction of battery electric vehicles (BEV) [14]. These vehicles, powered by renewably-generated electricity stored in batteries are a low carbon alternative to internal combustion engine vehicles (ICEVs) [15]. Even though the emissions from manufacturing BEVs are still greater, they can significantly reduce emissions per kilometre travelled as compared with conventional vehicles [11]. Therefore, they could have a significant impact in the reduction of urban air contamination and in transportation energy security [16].

The mileage range for most electric vehicles is around 96–145 km, and the batteries up to this range, are handy in size, compact and not too heavy. The mentioned vehicles are already technologically developed and have affordable prices. However, there is public distrust in the implementation of electric vehicles, mainly due to the longer mileage, faster refuelling and higher performance of conventional combustion engines. This opinion does not take into consideration that small electric vehicles have the capacity to more adequately and economically meet the mobility habits of a vast part of the population.

Complete information about road situations and traffic condition and accurate data on routes and driving cycles are of vital importance to determine the effectiveness of control strategies and to determine the advantages (if there are) of driving BEVs under real driving conditions.

The contributions of this study are the following: (1) the energy consumption of three different EVs was measured for different driving routes to analyse the effect of the road on the energy demand and to compare the results with the manufacturers' theoretical data. (2) The real energy consumption values were compared to the values obtained from the simulation software developed by Muneer et al. [14], with more accurate results than the ones shown in the dashboards of the vehicles. (3) The effect of the electricity generation mix, and thus policies related to the sustainability of different countries, on CO_2 was shown.

2. Literature Review

2.1. Decarbonisation of the Transport Sector

The problems related to the urban air pollution and energy security are accelerating the replacement of gasoline and diesel cars by battery electric vehicles. The 60% greenhouse gas emission reduction target established in the Transport White Paper 2011 [17] is one of the priorities of the European Commission [18], and many countries are establishing goals and bans related to the use and purchase of vehicles. Subsidies for BEV users, legislation and changing government policies and investments in charging infrastructures are some of the actions that different countries are taking towards this goal.

France, for example, with the objective of meeting their ambitious targets under the Paris climate accord, wants to stop the sale of diesel and petrol vehicles by 2040 [19]. Germany needs to reduce its greenhouse gas emissions by 40% of the emissions level in 1990 by 2020 to meet the German agenda for sustainable development and pollution control [20]. With this purpose, many cities and towns have introduced diesel bans [21]. As the purchasing power of the middle class increases, the number of vehicles in India is expected to increase notoriously in the near future, which will directly affect the pollution of many Indian cities, which are already some of the most contaminated in the world. Due to this fact, the Indian government wants to promote the purchase of electric vehicles and to ban the sale of vehicles that are not powered by electricity by 2030 [21]. In China, due to the increasing issues related to energy security and air pollution, the progress of electric vehicles (EV) is considered as national development strategy. Public acceptance and market insertion has increased considerably with the government's support. The electric vehicle industry in China became the largest EV market in the world in 2015, and it is growing and progressing greatly with the establishment of new policies [16].

Sweden has also set the objective of a fossil-fuel-free vehicle fleet by 2030 [22]. In Norway, the objective is to ban the sale of non-zero-emission vehicles by 2025, and because of generous policies supporting the increased use of electric vehicles, about the 40% of the cars sold in the country in 2016 were electric or hybrid vehicles [21]. There were only a few hundred BEVs in Norway in 2005, while in 2014, there were around 25,000 electric vehicles [10].

Britain is following a similar initiative. The UK Committee on Climate Change (CCC) claims that by 2020, 16% of the sales of new cars should be electric vehicles [14]. As part of the government's clean air plan, due to the rising levels of nitrogen oxide causing a major risk to public health, they plan to forbid new petrol and diesel cars and vans from 2040 [23]. The aim of the Scottish government is to achieve 100% non-fossil-fuel electricity by 2020 [24], and the disappearance of all non-electric vehicles from the road by 2050 [25]. For decarbonisation of the road, establishment of rapid charge points every at least 80 km has been approved, and, furthermore, the installation of home charging points has 100% funding [14].

The Scotland transport literature suggests that 94% of journeys in Scotland are less than 40 km, with an average distance of 12.1 km [26]. Therefore, most of the drives that take place in Scotland can be realised with an electric vehicle.

The objective of this research was to promote sustainable mobility and to offer a control strategy in Scotland. For this purpose, more than 50 electric vehicles' real-world operational data were monitored for a period of 4 years. The aim of reading and investigating this information was to analyse real working parameters and energy consumption, considering key operational features such as road types, speeds, distances and weather conditions, among others. The tests with the EVs were performed along the principal roads of Edinburgh, to simulate different road types, for different times of the day and weather conditions. The research also analysed the CO_2 emissions of different driving cycles and the effect of the energy generation mix in the emissions.

According to Morrissey, Weldon, and O'Mahony [27], the location of and confidence in the charging infrastructure affects the charging behaviour of BEVs; therefore, real and more accurate EV data will enhance the confidence of the users in these vehicles. A solid justification of this study is the necessity to increase the effectiveness of any control strategy which depends on accurate real-world data to support a real decrease of the environmental impact of the transportation sector.

2.2. Driving Cycle

There has been a notorious increase in interest in matters related to route planning, electric vehicle mileage range and speed optimisation and composition of vehicle fleets [28,29]. In this vein, battery electric vehicles are becoming more and more important, as they reinforce the introduction of sustainable delivery practices [30–32]. The previously mentioned growing government support, together with population increase and rising demand for zero-emission vehicles, will promote the improvement of BEVs in the future. The insertion of electric vehicles will not be smooth, since there are fears concerning their mileage [14]. Even though the energy and maintenance costs are much lower, EVs typically have less autonomy and payload limitations. The lithium-ion batteries used in these vehicles have a limited lifecycle and a specific charging and discharging pattern [30].

The actual energy demand of BEVs when they are operating on the streets is usually different from the manufacturer's specifications of the efficiency [33]. Legal measures are commonly based on defined and laboratory-controlled driving cycles, which are available just as summarised values. However, these values do not represent the local driving conditions and different driving cycles [34]. In day-to-day use, there are several factors, such as acceleration, deceleration, speed, driving duration and cruise times, among others, that determine the energy demand of electric vehicles for driving a certain route. It is not the same to drive in a city, where stopping is more frequent, than in a rural zone, and the person driving the vehicle also influences the previously mentioned parameters [20]. The driving cycle has a great effect on the energy demand [35].

On 1 September 2018, the new Worldwide Harmonised Light Vehicle Test Procedure (WLTP) approval cycle, more severe and realistic than the New European Driving Cycle (NEDC) used previously, became effective in Europe. The differences between the different approval cycles (NEDC,

EPA (Environmental Protection Agency) and WLTP) make it increasingly evident that the real autonomy of electric cars is not going to be the one announced by the manufacturer.

The "Mind the Gap" report by Transport and Environment describes the growing difference between official fuel consumption declared by manufacturers and the real consumption (and CO_2 emissions) of new cars, estimating that the difference between them is 25% [36].

Therefore, laboratory-controlled driving cycles are not sufficient for battery duration estimations, EV powertrain design or management of the batteries; transient and real driving cycles are vital to providing accurate information to electric vehicle users. The correlation between the energy demand and driving cycle of EVs has been reviewed in detail by Braun and Rid [20] and André [37]. The analysis and development of driving cycles for specific routes such as rural and urban areas would be of significant help in the design and optimisation of electric vehicles. Furthermore, this information would also contribute to increasing confidence in the economic, electricity grid and lifecycle studies conducted by governments and EV users.

3. Methodology

The objective of this research was to analyse real working parameters and energy consumption considering road types, speeds, distances and weather conditions. The research also analysed the CO_2 emissions of different driving cycles and the effect of the energy generation mix on the emissions. Figure 1 shows the main steps of the methodology.

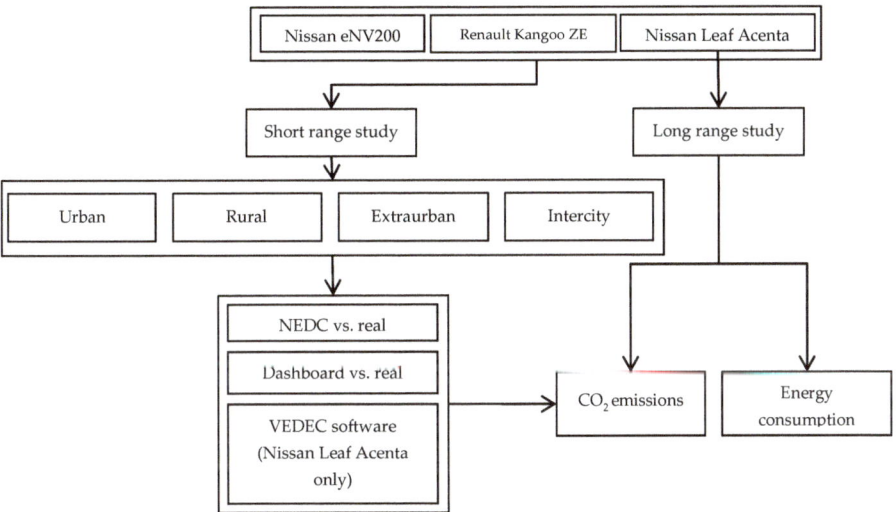

Figure 1. Flowchart illustrating the main steps of the adopted methodology.

3.1. Data Acquisition

The Edinburgh College established a green vehicle-leasing program in the 2012, with the objective of providing a more sustainable and low cost service to the staff. There are 10 electric cars in total, including the following manufacturers: Nissan, Renault, Mitsubishi and BMW electric cars, 4 Nissan vans and an Allied electric minibus in the college fleet available to rent. The most used cars and the Nissan van were selected for this study.

- Nissan Leaf Acenta.
- Renault Kangoo ZE.
- Nissan eNV200 combi comfort.

These vehicles were equipped and tracked to obtain more than 50 GPS and energy consumption data for 4 years. The table below (see Table 1) shows the technical information of each of the selected means of transportation.

Table 1. Technical information of the electric vehicles [38–40].

Technical Specifications	Nissan Leaf Acenta	Renault Kangoo ZE	Nissan Env200
Dimensions (m)	4.45 × 1.77 × 1.55	4.28 × 1.83 × 1.84	4.56 × 2.01 × 1.86
Max Power (kW/rpm)	80	44	80
Acceleration from 0 to 100 km/h (sec)	11.50	22.30	14
Autonomy (km)	199	165	167

The development of driving cycles involves [37]:

- Recording driving conditions using one or several instrumented vehicles driven for their normal purposes.
- Analysing these data in order to describe or characterise these conditions.
- Developing one or more representative cycles for the recorded conditions.

The drive cycle data were collected live for the experiment using the vehicles' own electronic control unit (ECU) memory and the CANbus diagnostic port, using a LUFII ELM327 V1.5 mini-wireless OBD2 diagnostic scan tool that was retrofitted.

3.2. Experimental Analysis

For the experimental analysis of this research work, rural, urban and intercity routes in the Edinburgh area were selected to monitor and record the energy consumption and speed of the EVs at various times of the day to obtain economic, environmental and energy performance parameters.

Each route type had specific characteristics. Urban routes are normally roads entering cities and towns, such as the artery into Edinburgh from the south. It is common to stop and start the car several times due to the traffic flow and speed restrictions. Extra-urban routes usually have larger cruise periods, since these roads are partially rural. The traffic flow determines the speed of the vehicle. The rural route in this test started in the Sighthill campus and ended in the Midlothian campus. Although a rural route, there were some urban areas with slower speed sections.

The tests were conducted using the car chasing technique; according to Chaari and Ballot [41] and confirmed by Andre [42], this method:

- Decreases the possibility of manipulation of the driver's performance.
- Assures the usual driver uses the vehicle.

The drivers of the vehicles were staff of the Edinburgh College of different genders and areas for a close representation of the drivers in the region, and in all cases the drivers respected legal limitations on the road.

The only way of knowing the remaining power, and therefore the available distance an electric vehicle has is the display of the dashboard. Thus, the accuracy of this information is of vital importance for journey planning.

In order to compare the information that the displays showed with real values, data loggers were installed in the car with the objective of reading and storing the parameters from the vehicles control area network (CAN) where the sensors were located. CAN bus information was recorded every second, while the GPS position was logged every 5 s.

This research contemplated four different studies, which are detailed in the following section and summarised in Table 2. The experimental trials were carried out in summer and winter climate situations to obtain greater accuracy in the results and to allow data study on both air conditioning and heater usage and the auxiliary energy consumption used for heating the seats or defrosting before driving, among other features, that the driver would use in a normal situation [43].

Table 2. Characteristics and objectives of the test.

Experiment		Vehicle Used	Route	Aim of the Study
a	80 km route for 3 BEVs	• Nissan Leaf • Renault Kangoo ZE • Nissan eNV200	• Mixed route	• NEDC value vs. dashboard value • Dashboard value vs. real value • CO_2 emissions analysis
b	Urban, rural, mixed combined and intercity cycles	• Nissan Leaf	• Urban • Rural • Intercity • Extra-urban	• NEDC value vs. real value • Dashboard value vs. real value • CO_2 emissions analysis
c	VEDEC simulation software	• Nissan Leaf	• Urban • Rural • Extra-urban	• VEDEC software estimation vs. real value
d	Long-range driving test	• Nissan Leaf	• Long distance (mixed routes)	• Performance, energy use and CO_2 emissions analysis and comparison

(a) 80 kilometre route for 3 BEVs

This study compared the energy consumption of the previously mentioned three electric vehicles covering the same distance on the same route segment. For this purpose, the remaining distance shown in the dashboard was compared with the New European Drive Cycle (NEDC) values (these values are given by the car manufacturer to control the accuracy of the vehicle's electronic control), and with the real value provided by the data loggers.

The route selected for these tests was 80 km long, it included both urban and rural zones, and the topography of the road included diverse driving styles to ensure real driving conditions. As the trials were conducted at various times of the day, they represented realistic traffic conditions [44]. The figure below shows the route used in the first study (see Figure 2).

The trials were conducted in both winter and summer, when the external temperature was between −2 and 2 °C and between 16 and 20 °C, respectively. The internal temperature in the car was set to 20 °C for the driver's comfort during the driving tests. According to Alahmer et al. [45], the temperature in the interior of the car affects the driver's behaviour, thus, in order to reduce the user's stress and fatigue and ensure comfort, the internal temperature was set to 20 °C.

(b) Urban, rural, mixed combined and intercity cycles

This test objective was to compare the dashboard values, NEDC information and real driving data for the same car in urban, rural, extra-urban and intercity routes, in order to analyse the effects of the driving mode and different weather conditions on the energy consumption.

The vehicle used for the analysis of different driving patterns was the Nissan Leaf model [46,47]. In this case, the only information given to the driver was that the vehicle needed to get from one location to another one, without a planned route. This way, it was possible to consider all driving conditions without a specific pattern. In all the tests, legal limitations and traffic conditions were followed.

The trials were conducted both in summer time, when the external temperature was between 16 and 20 °C, and in winter time, when the external temperature was between −2 and 2 °C, to analyse the effect of temperature on the vehicle's behaviour and ensure a higher accuracy in the results. The interior temperature in the car was 20 °C throughout the entire test, using the heater in winter and cooling ventilation in the summer time.

Auxiliary equipment such as radio, seat heater or interior lighting was used when necessary depending on the journey, since this would be closer to real world conditions and it is senseless to conserve auxiliary energy [43].

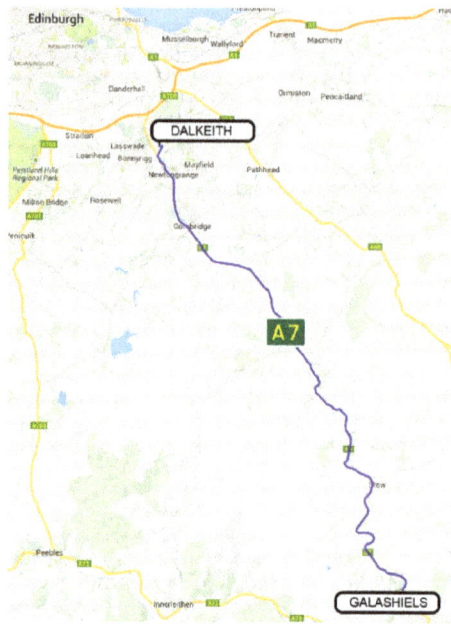

Figure 2. Experimental commute route (Google Maps, 2018).

(c) VEDEC simulation software

For the proper development of electric vehicles, proper simulation of the driving cycle is becoming more important. Muneer et al. [14], based on the study of Rubin and Davidson [48], developed a simulation software called Vehicle Dynamics and Energy Consumption (VEDEC). This software is written in VBA from Microsoft Excel Software and uses dynamic equations for the calculations. It can estimate the energy and power needs for any vehicle when driving, and it also calculates the available energy that a car can gain from regenerative braking, compared to the same vehicle without that system. The software evaluates the differences in energy consumption in different driving modes, such as acceleration, cruise and gradient-climbing, logging topography maps to the on-board altimeter. Note that VEDEC software was developed by the present team and was one of the constituent elements of Milligan's doctoral research program [49]. The software has been extensively validated using the measured energy consumption of test vehicles, which included a Renault Kangoo ZE and a Nissan Leaf.

The lack of confidence in the theoretical autonomy of EVs has led to the development of various software and/or models to estimate the available mileage range of these vehicles, such as the analytical model developed by Wu et al. [50]. In their study, they first presented a system which can collect in-use EV data and vehicle driving data. Approximately 5 months of EV data were collected, and these data were used to analyse both EV performance and driver behaviour. Fiori et al. [51] developed the Virginia Tech Comprehensive Power-based EV Energy consumption Model (VT-CPEM) to estimate the driving parameters of EVs, and Hayes et al. [52] presented a simplified EV model to quantify the impact of battery degradation with time and vehicle auxiliary loads for heating, ventilation, and air conditioning (HVAC) on the total vehicle energy consumption.

The third part of this study compared the real driving parameters obtained from the data loggers with the estimations of this software tool. For these estimations, the program used current experimental inputs to determine real parameters for a specific location. The simulations in this research were conducted with a 95% efficiency for the motor and 60% for the regeneration.

For the driving test, a Nissan Leaf model was used; thus, the tools to measure and read the data were logged in the car. The recorded parameters were the acceleration and deceleration rates, altitude to estimate the inclination difference, speed, distance and duration of the trip. With these parameters, the driving cycles were determined for comparison with the estimated values from the software.

The test route this time included urban, extra-urban and rural areas, with the aim of identifying factors determined by acceleration and speed that were unique for a specific route type. For this purpose, the car was charged to 100% energy level and driven to the destination, recording the energy consumption. Once arrived at the destination, the car was recharged. Distance, journey duration and charge duration were also recorded. Figure 3 shows the mobility activity (%) of the test vehicle on each type of road, showing the percentages of time the vehicle was accelerating, decelerating, at a constant speed (cruise) or stationary. The results were different depending on the road type. In the extra-urban route for example, due to the dual-carriageway nature of the road, higher and more constant speeds were reached, and therefore the cruise section was larger. The amount of time spent in each mobility activity directly affected the energy consumption.

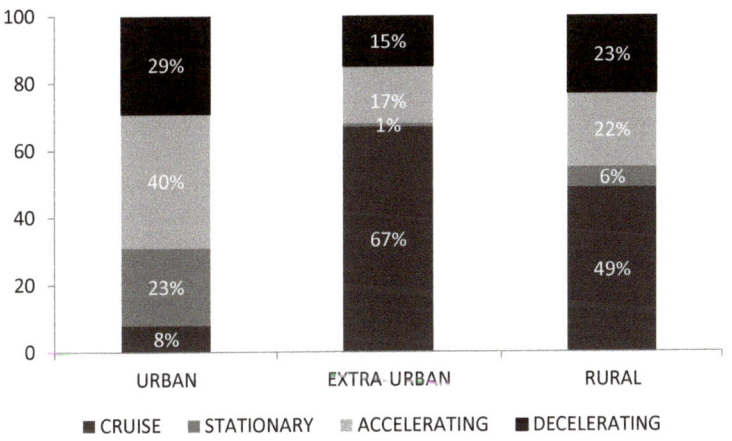

Figure 3. Mobility activity in different routes.

(d) Long-range driving test

In the previous tests, all the journeys were within the driveable range of an EV. However, the limited autonomy and poor charging infrastructure are the main drawbacks of the electric vehicles. Therefore, the aim of the last test of this study was to analyse the performance, energy consumption and CO_2 emissions of these vehicles over long distances. These data will help in the improvement of the batteries and in the development of the rapid charger infrastructure.

The Nissan Leaf Acenta electric vehicle was selected to record the long-range test using the data recording equipment.

The journey was conducted over several days through the A701-M74-M6-M5 main route from Edinburgh to Bristol, which includes a rapid chargers network known as the "Ecotricity electric highway" [53]. The trip was divided into different sectors, so the effect and dependence of the elevation and road conditions could also be studied, and the energy consumption on different terrains could be compared with internal combustion engine (ICE) vehicles. Figure 4 shows the distance of the test.

Figure 4. Long-range route (Google Maps, 2018).

3.3. CO₂ Emissions Calculation

Electric vehicles' GHG emissions are nil when driving, however, there are emissions related to the generation of the electricity that drives the battery that need to be considered. The quantity of the emissions depends on the energy source used in the electricity generation. The dependence on fossil fuels is being reduced with the help of legislation and new government policies.

In Scotland, with the aim of achieving a complete decarbonisation of the roads by the 2050, the Scottish government has invested not only in a sustainable fleet of electric vehicles, but also in the renewable energy sector. The equivalent grid carbon intensity in Scotland in 2014 was 196 gCO_2/kWh, and it was reduced to 151 gCO_2/kWh by 2015 [54]. As indicated in Table 3, 42.4% of the power in Scotland is generated using renewable energy sources. In the UK, the dependence on fossil fuel is higher and only the 24.5% of the electricity comes from renewable energy sources; thus, the equivalent carbon emission is 458 gCO_2/kWh [55]. Note that taking into account the energy losses between the source of electricity and the charging station, the efficiency of the charging station and the efficiency of the batteries, a total loss of 22% has been considered [14,56].

Table 3. Electricity generation by fuel type (Gov. UK, 2016).

Energy Sources (%)	Scotland		UK	
	2015	2016	2015	2016
Coal	16.10	3.90	22.40	9.00
Gas	3.70	6.80	29.50	42.20
Nuclear	34.60	42.80	20.70	21.10
Renewables	42.40	42.90	24.60	24.50
Oil and Other	3.20	3.60	2.80	3.20

4. Results

4.1. 80 Kilometre Route for Three BEVs

The elevation of the route is shown in Figure 5, and the driving cycle of the tests can be seen in Figure 6.

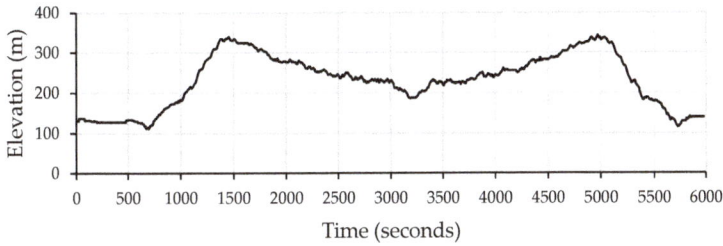

Figure 5. 80 km monitoring cycle—elevation—from Midlothian to Galashiels.

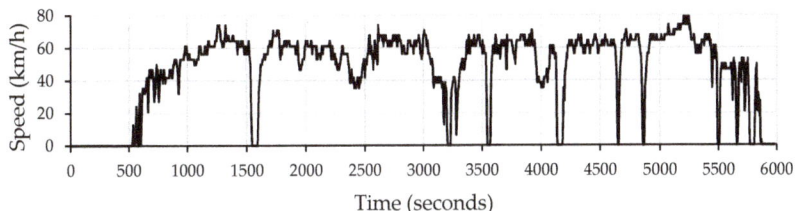

Figure 6. 80 km monitoring cycle—speed—from Midlothian to Galashiels.

Table 4 compares the NEDC values of each vehicle, provided by the manufacturers, with the estimation of the available distance shown in the dashboard and the real values obtained from the 80 km driving test performed with each electric vehicle. In the four cases, the dashboard showed a lower autonomy and the real driveable range was even lower. This fact only enhances the distrust of users of EVs, due to concerns about the possibility of not reaching their destinations.

Table 4. Range comparison between NEDC values, dashboard estimations and real values.

Vehicle Used	Nissan Leaf	Renault Kangoo ZE	Nissan Env200 Unladen	Nissan Env200 (Laden 500 kg)
		Summer		
Max. range acc. NEDC (km)	199.60	170.60	169.00	169.00
Estimated range in dashboard (km)	149.70	112.70	not available (n.a.)	n.a.
Real value (km to depletion)	130.50	109.90	n.a.	n.a.
Energy efficiency (km/kWh)	5.92	5.15	n.a.	n.a.
Estimated vs. NEDC (%)	−25.00	−33.96	n.a.	n.a.
Real vs. estimated (%)	−12.80	−2.43	n.a.	n.a.
CO_2 emissions/80 km (kg)	2.62	3.00	n.a.	n.a.
		Winter		
Max. range acc. NEDC (km)	199.60	170.60	169.00	169.00
Estimated range in dashboard (km)	120.70	112.70	112.70	138.40
Real value (km to depletion)	123.10	95.40	109.40	114.70
Energy efficiency (km/kWh)	5.89	4.83	5.75	5.99
Estimated vs. NEDC (%)	−39.52	−33.96	−33.33	−18.10
Real vs. estimated (%)	2.00	−15.29	−2.86	−17.14
CO_2 emissions/80 km (kg)	2.05	2.50	2.10	2.02

The vehicles' estimations visible in the dashboard were around 30% lower than the theoretical values of the vehicles. In the case of the Nissan Leaf, the difference between both parameters was larger in the winter period, so the vehicle is sensitive to the external temperature and auxiliary energy consumption. When comparing the dashboard values with the real autonomy of the Nissan Leaf, the results showed that the driveable distance in winter was higher than the autonomy shown in the dashboard, which confirms the effect of the external weather conditions on the vehicle's sensors. In the case of the Renault Kangoo ZE, the difference between the NEDC and dashboard values was the same during winter and summer; thus, the vehicle's electronics are not sensitive to weather conditions. However, the real autonomy of the car was highly affected by the external temperature. The difference between the dashboard and real values was 2.43% in summer and 15.29% in winter.

The energy efficiency represents the distance travelled using a given amount of energy. The higher the efficiency, the lower the energy consumption, and thus the lower the CO_2 emissions. The results in Table 4 show that the Nissan Leaf vehicle had the lowest CO_2 emissions among all the cars in the research. However, the weather conditions had an effect on the emissions.

Figure 7 illustrates the comparative values of the NEDC, predicted range and real driving cycle for the Nissan Leaf during the winter period. Due to the lower external temperature when operating in winter, the results did reflect a decreased range capability.

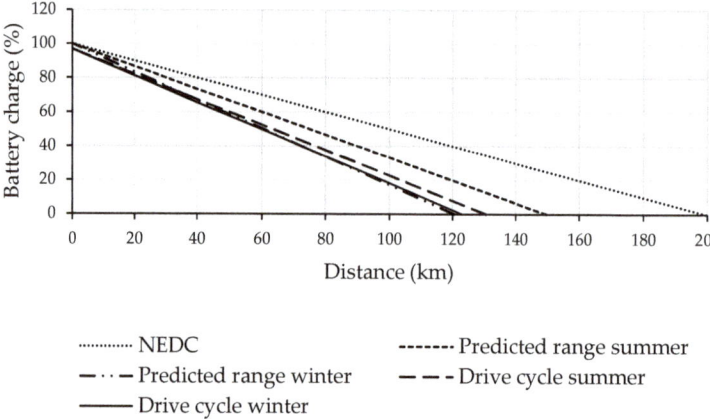

Figure 7. Nissan leaf range comparison.

Figure 8 shows the consumption of the auxiliary energy for heating the Nissan Leaf vehicle during the winter trials to defrost the car prior to the driving test. The initial energy demand reached a peak of 4.25 kW and an additional 3.75 kW for about 6 min. These affected the mileage range achievable during winter compared to the summer period. Fiori et al. [51] also found similar results regarding the effect of auxiliary systems.

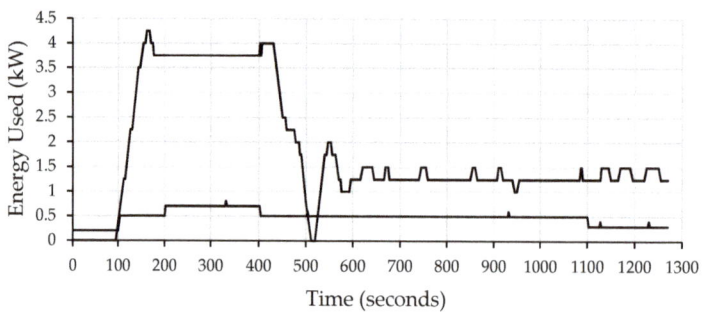

Figure 8. Auxiliary energy demand during the first 21 min.

The driveable length of the vehicles is an indicator of the energy efficiency and it is directly related to the CO_2 emissions of the vehicles. Therefore, as exposed in Table 4, the CO_2 emissions were higher when the efficiency decreases. Consequently, the emanation of hazardous gases was higher during winter periods.

At this point, it is important to mention that the maximum availability of the battery when it was fully charged and the display in the dashboard indicated 100% was not real. Before the start of each trip, the vehicle was completely charged, indicated as 100% in the display of the dashboard, and checking that the charging post showed that further charge was not possible. These data was verified using the vehicle's electronic control (ECU) unit and CANbus through the diagnostic port with a hardware plugin. The results showed that the ECU was recording a lower value of 97% for the state of charge (SOC) when fully charged, instead of 100% as expected. The experiment was repeated in many Nissan Leaf cars, and the values mentioned were continuously repeated. The maximum value that could be achieved was 22 kWh, instead of 24 kWh as stated by the manufacturers, giving a 2 kWh margin for safe battery control and eliminating the risk of overcharging, irrespective of the charging source that the car is plugged into.

The variation between the maximum capacity and the energy availability can be caused by different facts. According to Lam [57], it is related to the temperature and it will be affected throughout the battery's life. However, when the battery is being used with higher currents, the available capacity can also be lower.

There is a need for a further research in the modelling battery recharging and discharging processes in a tractable way [30]. According to Sauer [58], the battery's available energy should be measured as an electrical charge rather than as energy. Bergveld [59] reinforced this fact, claiming that the SOC should be the ratio between the electrical charge in the battery and the maximum charge it can have. The fraction between the electric current of the battery and the maximum capacity can define the instantaneous variation of the SOC [60]. The battery's terminal voltage and current can vary depending on the power pattern of each drive cycle [61]. The battery's power output depends on the terminal voltage and current [62].

4.2. Urban, Rural, Mixed Combined and Intercity Cycles

The figures below show the elevation and drive cycle of the extra-urban route. Figure 9 illustrates the elevation and duration of the extra-urban route test. Figure 10 shows the return drive cycle, in which the speed was approximately 50–65 km/h on the urban roads and 80–96 km/h on the dual-carriage roads for the same road type, where the energy recovery was applicable in the start and stop of the test and will therefore be insignificant.

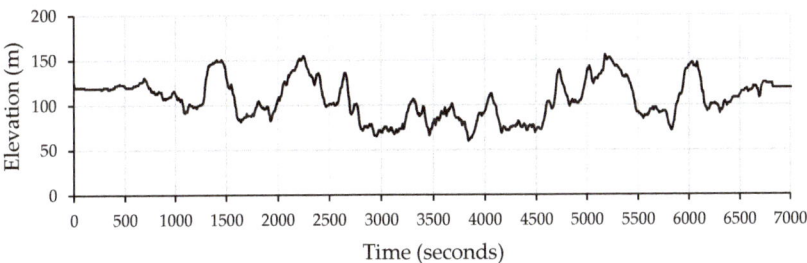

Figure 9. Elevation of the extra-urban route (m).

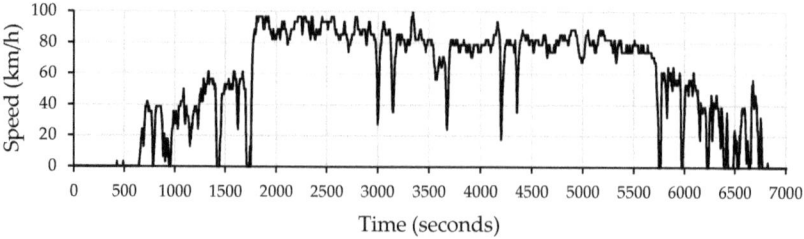

Figure 10. Speed of the vehicle during the extra-urban route.

The second test of the study analysed the difference between the real driving cycle and the estimated range shown in the dashboard display and NEDC data from the manufacturers for urban, rural, intercity and extra-urban routes.

These tests were conducted in both summer and winter with a Nissan Leaf vehicle, the external temperature being between −2 and 2 °C in winter and between 16 and 20 °C in summer. The results indicate that the real energy efficiency of the vehicle was considerably lower than the estimations, especially during the winter (see Table 5).

Table 5. Nissan Leaf range comparison for different routes.

Route	Urban	Rural	Intercity	Extra-Urban
Summer				
Max. range acc. NEDC (km)	199.60	199.60	199.60	199.60
Estimated range in dashboard (km)	149.70	149.70	136.80	149.70
Real value to depletion (km)	143.20	130.50	118.40	135.20
Energy efficiency (km/kWh)	6.50	5.92	5.38	5.63
Estimated vs. NEDC (%)	−25.00	−25.00	−31.45	−25.00
Real vs. estimated (%)	−4.30	−12.80	−13.41	−9.68
CO_2 emissions/80 km (kg)	2.38	2.62	2.88	2.75
Winter				
Max. range acc. NEDC (km)	199.60	199.60	199.60	199.60
Estimated range in dashboard (km)	138.40	120.70	148.10	133.60
Real value to depletion (km)	114.30	123.10	115.20	111.40
Energy efficiency (km/kWh)	6.00	5.90	5.30	5.60
Estimated vs. NEDC (%)	−30.65	−39.52%	−25.81	−33.06
Real vs. estimated (%)	−17.44	2.00	−22.17	−16.63
CO_2 emissions/80 km (kg)	2.58	2.62	2.92	2.77

The values in the dashboard were significantly lower than the theoretical NEDC values provided by the manufacturer. The difference was higher in winter. However, the real driveable range of the Nissan Leaf in different routes was even lower than the estimation of the vehicle. Figure 11 shows the variance between the NEDC, estimated and real driving discharge periods for the extra-urban route. The reduction in the distance available in the real driving was significant compared to the NEDC values.

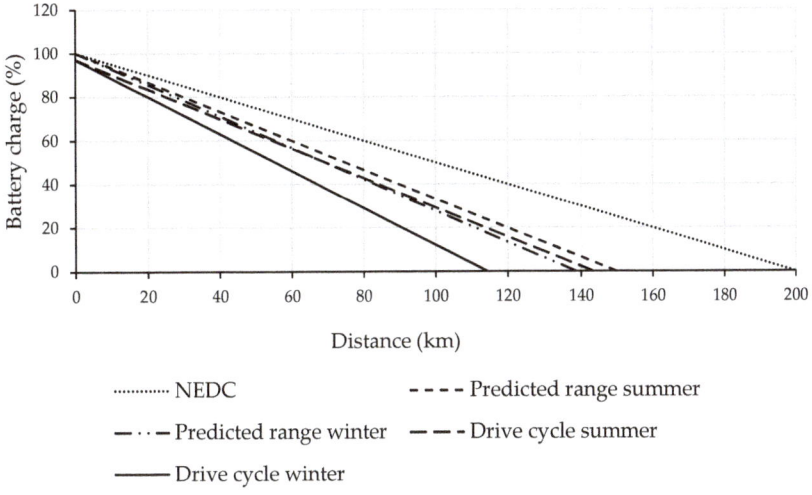

Figure 11. Discharge comparison for extra-urban route.

During summer, the autonomy of the vehicle was 143.2 km, but in winter, the available energy was only enough for 114.2 km. This means a reduction of 20.25% in range. The route was the same in both seasons, and the same driver drove the vehicle in equal conditions. The only variables in the tests were the weather and climatic conditions such as temperature, rain and wind. This fact could have a negative impact on the driver and vehicle reaching their journey's end. The difference in the driving patterns for each route, and the effect of the drive cycle in the battery range, confirmed the research led by Neubauer, Brooker, and Wood [63].

Regarding the energy efficiency and CO_2 emissions of the vehicle on different types of road, the results coincided with many other studies in the fact that the efficiency of the EVs was higher in urban areas compared to rural, intercity and extra-urban areas [50,51]. Again, the temperature had a negative effect in the results. It is also worth mentioning that on roads where the speed was maintained constantly during a period, as, for example, on extra-urban and intercity routes, the energy recovery will be negligible, since there will not be regeneration. In urban roads, by contrast, the start–stop process of the car was more frequent and therefore the CO_2 emissions were lower.

4.3. VEDEC Simulation Software

The third test of the study compared the real drive cycles with simulations conducted using the VEDEC software in order to verify the accuracy of the simulation tool on different route types. For this test, a Nissan Leaf vehicle was driven through urban, rural and extra-urban roads.

A critical parameter when simulating the energy consumption and speed is the rolling friction of the vehicle. The greater the friction, the lower the distance travelled will be, since the energy demand to rotate the wheels will be higher. The right hand side of Equation (1a) has components of energy that include, from left to right, tyre friction, ill climbing, wind drag and change in kinetic energy [14]. Assuming travel on a level road (Equation (1b)) and making μ the subject in Equation (1b), Equation (1c) [64] was used to estimate the friction coefficient for the vehicles in this test, with a value of 0.0269; thus, the coefficient utilised for the simulation was between 0.025 and 0.03 depending on the surface material of the road and the tyre pressure [43]. Note that the authors used the cars mentioned in Figure 1. The cars were produced in the years 2011 (Nissan) and 2012 (Renault). The value of the coefficient of friction quoted was based on information obtained from the manufacturers. Those values were found to be in agreement with our experimental results.

$$E = \left[\mu mg\cos\theta + mg\sin\theta + \frac{1}{4}C_d A\rho(v_f^2 + v_i^2)\right]\Delta d + \frac{1}{2}m(v_f^2 - v_i^2) \tag{1a}$$

$$E = \left[\mu mg + \frac{1}{4}C_d A\rho(v_f^2 + v_i^2)\right]\Delta d + \frac{1}{2}m(v_f^2 - v_i^2) \tag{1b}$$

$$\mu = \frac{\frac{E - \frac{1}{2}m(v_f^2 - v_i^2)}{\Delta d} - \frac{1}{4}C_d A\rho(v_f^2 + v_i^2)}{mg} \tag{1c}$$

C_d: coefficient of drag; A: frontal area; ρ: air density; m: mass; g: acceleration gravity; v_f: final velocity; v_i: initial velocity; E: energy; Δd: distance.

Vehicle manufacturers will recommend the lowest friction coefficient for the tyres, since it will affect in the mileage range of the vehicles.

Table 6 shows the energy consumption and speed estimated values of the software compared to the real drive cycles for urban, extra-urban and rural routes. In the case of the urban route, the energy demand estimation was 7.30% higher than the real energy demand and the speed was 2.12% lower than the real value. The energy meter norm for APT control units was: Active (kWh) = BS EN 50470-3 (Class B ± 1%). Thus, the estimated results of the simulation software are reliable.

Table 6. Energy, distance and speed—overview.

Energy Consumption and Speed	Urban			Extra-Urban			Rural		
	Simulation	Real	Variation	Simulation	Real	Variation	Simulation	Real	Variation
E used (kWh)	2.67	n.a.	n.a.	6.73	n.a.	n.a.	5.64	n.a.	n.a.
E regen (kWh)	0.34	n.a.	n.a.	0.32	n.a.	n.a.	0.49	n.a.	n.a.
E tot (kWh)	2.33	2.16	7.30%	642	6.72	−4.74%	5.15	5.28	−2.56%
Duration (sec)	1196	1196	n.a.	1729	1729	n.a.	2381	2381	n.a.
Distance (m)	11,909	12,162	n.a.	30,095	29,008	n.a.	27,836	27,041	n.a.
Average speed (km/h)	35.85	36.61	−2.12%	62.66	60.40	3.61%	42.09	40.89	2.86%

The accuracy of the estimation values obtained with the simulation software in the case of the extra-urban and rural routes was much higher. The energy demand estimation was 4.74% and 2.56% lower for the extra-urban and rural roads, respectively, while the speed was 3.61% and 2.86% higher in each case. Both routes had longer cruise periods than the urban route, especially in the case of the extra-urban route which had a dual-carriageway nature. This fact resulted in long high-speed periods which increased the energy demand even though the journey time was 25% lower than in the rural route. Both routes showed similar periods of acceleration and deceleration. The rural route had the highest accuracy in the estimated values.

Fiori et al. [51] developed an energy consumption estimation model which produces an average error of 5.9% relative to empirical data. Wu et al. [50] proposed an analytical EV power estimation model with a mean absolute error (MAE) of 15.6%.

Table 7 illustrates the comparison between the route distances shown on the vehicle's dashboard with the estimated values of simulation software and the distances taken from proprietary GBmapometer and Google maps sources. The objective of this comparison was to verify the accuracy of the vehicle's parameters and to ensure the UK legislative purposes for any speedometer were fulfilled [14].

Table 7. Travelled distance comparison and error.

Test Number	Vehicle Odometer Reading (m)	Simulation		GBmapometer		Google Maps	
		Distance (m)	Conformity (%)	Distance (m)	Conformity (%)	Distance (m)	Conformity (%)
1	27,842.00	28,031.74	99.30	27,440	98.60	27,358.80	98.30
2	29,766.50	30,872.73	96.40	29,590	99.40	29,611.90	99.50
3	11,909.00	12,161.70	97.90	12,350	96.40	11,748.20	98.60
4	12,231.00	12,147.66	99.30	12,340	99.10	11,748.20	96.10
5	30,095.00	29,008.39	96.40	29,680	98.60	29,611.90	98.40
6	27,835.70	27,040.70	97.10	27,390	98.40	27,358.80	98.30
Total			97.70		98.40		98.20

The distance read by the vehicle was very similar to the GBmapometer and Google maps distances, with less than 2% variation. The accuracy of the measurements of these tools was dependent on the terrestrial reference system (TRS). The distance values estimated using the simulation software were within 1% of the distances taken by GBmapometer and Google maps. Therefore, the software's results are reliable.

4.4. Long-Range Driving Test

The long-range driving test was carried out using the Nissan Leaf vehicle from Edinburgh to Bristol. Table 8 shows the distance travelled and energy consumption between each charging point (sector). The CO_2 emissions of each sector were estimated using both Scottish and UK grid carbon intensity figures and compared to the emissions of an equivalent internal combustion engine (ICE) vehicle.

Table 8. Travelled distance, energy consumption and CO_2 emissions.

Sector	Distance (km)	Energy Consumption (kwh)	Energy Efficiency (km/kwh)	CO_2e Scotland (kg)	CO_2e UK (kg)
1	92.61	17.50	5.29	3.39	10.28
2	68.87	13.90	4.95	2.69	8.16
3	61.16	10.30	5.94	1.99	6.05
4	83.33	13.70	6.08	2.65	8.04
5	83.72	15.20	5.51	2.94	8.93
6	74.75	13.80	5.42	2.67	8.10
7	105.53	15.20	6.94	2.94	8.93
Total	569.97	99.60	5.72	19.28	58.49

The weather and road conditions (see Table 9) had an effect on the efficiency of the electric vehicle. It was noticed that the energy consumption rate was greater at higher driving speeds.

Table 9. Road conditions.

Sector	Road Conditions
1	Poor weather, inclination
2	Poor weather, 97 km/h speed maximum
3	Poor weather, 81 km/h speed maximum
4	Motorway section
5	Motorway section
6	Motorway section
7	Motorway section, behind a good large vehicle

Table 10 shows the consumption and emission values of the Nissan Leaf vehicle and an equivalent internal combustion engine (ICE) vehicle for the same route. The similar-sized passenger vehicle used for the comparison was the Ford Focus 1.6.

Table 10. Electric vehicle vs. internal combustion engine.

Type of Vehicle	Distance (km)	Consumption	CO_2 at Tailpipe (kg)	CO_2e Scotland (kg)	CO_2e UK (kg)
Electric Vehicle	569.97	Energy (kWh)	nil	Energy generation	Energy generation
		99.60		19.28	58.49
Internal Combustion Engine	569.97	Fuel (l)	77.50	Fuel production	Fuel production
		46		9.17	27.80

The emissions of an electric vehicle depend on the grid mix used to generate the electricity that powers the battery. The grid mix values used for these estimations were mentioned in the Section 3.3, 0.151 kg/kWh in Scotland and 0.458 kg/kWh in the UK. The CO_2 emissions at tailpipe are nil. According to Lane [65], the tailpipe emissions of the equivalent ICE vehicle are 0.136 kg/km, therefore, in this route, the total CO_2 emissions would ascend to 77.5 kg. However, for a proper comparison of both vehicles, it is essential to consider the CO_2 emissions of the production of the fuel used to power the car. Both the Scotland and UK values have been used for this estimation.

The results prove that the switch from internal combustion engines to electric vehicles is essential to achieve a more sustainable transport. Nevertheless, electric vehicles are not free from emissions, and the numbers shown in Figure 12 show their dependency on the grid. Scotland is very focused on the achievement of 100% non-fossil-fuel electricity by 2020. Currently, when an electric vehicle is charged in Scotland, it emits 5.7 times less CO_2 than an ICE vehicle. In England, the dependency on fossil fuels is much larger, and it reduces 2.3 times the emissions of a conventional vehicle.

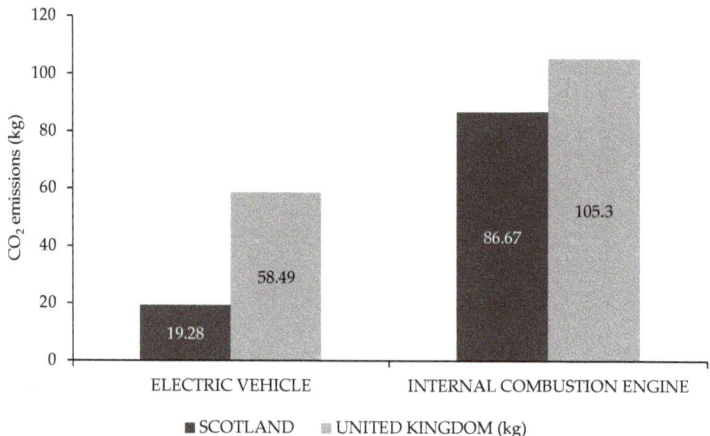

Figure 12. Scotland vs. UK CO_2 emissions.

5. Conclusions

Even though the mileage range of the electric vehicles in this study was around 160 km, four times lower than the range of a conventional fuel vehicle, it has been proven that the 94% of the journeys that take place in Scotland are less than 40 km. The acceptance and confidence of users in electric vehicles is still low. One of the main reasons for this is the lack of reliable information about the achievable range and performance of an electric vehicle.

The results of this study indicate that the data provided by the manufacturers (NEDC) and the values shown in the vehicle's dashboard are inaccurate, and far from being accurate. The values shown in the dashboards of the vehicles were, in most cases, 30% lower than the theoretical values according to the manufacturers (NEDC). The difference between the dashboard values and real mileage range was as high as 13% in some cases. The gap between the estimated values and the real range of the vehicle was larger during the winter, with differences up to 17% in some cases. The Nissan Leaf vehicle had 20% less autonomy during winter for the same extra-urban route. This explains the fact that there is a decrease in use of the electric vehicles in winter.

The results of the simulation software were much more accurate than the vehicle's information on the display; thus, it is possible to have more accurate information. The difference between real and estimated values ranged between +7.9% and −4.5%. This reduces the risk of battery depletion and proves that it is possible to have a real prediction of the vehicles' mileage range using data from real driving patterns. The dashboard information of the vehicles needs to be modified and adapted to real values to gain accuracy in the distance and energy availability, and thereby increase the confidence of electric vehicle users.

Therefore, it is necessary to update the performance measurement systems of the different vehicles, since the car manufacturers are cheating the policy makers, manipulating the tests and producing vehicles that only reach their regulatory objectives during tests and not on the road, where the fuel is burned and emissions occur. Regulatory pressure to reduce CO_2 emissions from new cars, significant tax exemptions for cars with low CO_2 test figures and high oil prices have increased the incentive for car manufacturers to manipulate test results. This manipulation will continue unless policy makers act to change things.

Regarding the CO_2 emissions of electric vehicles, the results showed that these were lower than the emissions of internal combustion engines. However, even though the emissions in the tailpipe are nil, there is a clear dependency on the electricity grid. Many countries, with the aim of achieving the 60% GHG emission reduction target, are establishing policies and bans to reduce the use of diesel and petrol vehicles. There are subsidies to promote the use of electric vehicles. However, as the results of this study show, the emissions of EVs depend on the electricity generation mix; thus, it is essential to focus on the electricity generation mix and to reinforce green energy sources in order to achieve a sustainable transport system. Scotland's electric infrastructure has augmented the contribution of renewable energy sources from 32% in 2013 to 42.9% in 2015. Therefore, currently, an electric vehicle in Scotland emits three times less than a vehicle charged in another location in the UK for the same route.

Electric vehicles are more sustainable and less hazardous than internal combustion vehicles. However, it is vital to focus on the power generation sources in order to find a solution to urban air pollution and energy-security-related problems.

Further work will be needed to strengthen the present work by collecting more data on all types of routes—urban, rural and mixed, combined and intercity. The present study was not truly exhaustive, so further data will bring into play a combination of varying speeds, road altitudes and driver behaviour.

Author Contributions: Conceptualization, R.M. and T.M.; methodology, R.M. and T.M.; software, R.M. and T.M.; validation, R.M. and T.M.; writing-original draft preparation, S.E. and E.J.G.; writing-review & editing, E.J.G.

Funding: This research received no external funding.

Conflicts of Interest: The authors declare no conflict of interest.

References

1. World Oil Outlook 2017. Available online: https://www.opec.org/opec_web/flipbook/WOO2017/WOO2017/assets/common/downloads/WOO%202017.pdf (accessed on 14 May 2019).
2. Liu, L.; Chen, R.-C. A novel passenger flow prediction model using deep learning methods. *Transp. Res. Part C Emerg. Technol.* **2017**, *84*, 74–91. [CrossRef]
3. Dulebenets, M.A. A novel memetic algorithm with a deterministic parameter control for efficient berth scheduling at marine container terminals. *Marit. Bus. Rev.* **2017**, *2*, 302–330. [CrossRef]
4. Dulebenets, M.A. A comprehensive multi-objective optimization model for the vessel scheduling problem in liner shipping. *Int. J. Prod. Econ.* **2018**, *196*, 293–318. [CrossRef]
5. UNCTAD. *Review of Maritime Transport 2018*; United Nations Publications: New York, NY, USA, 2018.
6. Akturk, M.S.; Atamtürk, A.; Gurel, S. Aircraft rescheduling with cruise speed control. *Oper. Res.* **2014**, *62*, 829–845. [CrossRef]

7. Jalalian, M.; Gholami, S.; Ramezanian, R. Analyzing the trade-off between CO_2 emissions and passenger service level in the airline industry: Mathematical modeling and constructive heuristic. *J. Clean. Prod.* **2019**, *206*, 251–266. [CrossRef]
8. Wagenaar, J.; Kroon, L.; Fragkos, I. Rolling stock rescheduling in passenger railway transportation using dead-heading trips and adjusted passenger demand. *Transp. Res. Part B Methodol.* **2017**, *101*, 140–161. [CrossRef]
9. European Petroleum Refiners Association AISBL. Available online: https://www.fuelseurope.eu/knowledge/refining-in-europe/fuelling-the-eu/transport/road-transport/ (accessed on 15 July 2019).
10. Holtsmark, B.; Skonhoft, A. The Norwegian support and subsidy policy of electric cars. Should it be adopted by other countries? *Environ. Sci. Policy* **2014**, *42*, 160–168. [CrossRef]
11. Bauer, G. The impact of battery electric vehicles on vehicle purchase and driving behavior in Norway. *Transp. Res. Part D Transp. Environ.* **2018**, *58*, 239–258. [CrossRef]
12. European Environment Agency. Energy Efficiency and Energy Consumption in the Transport Sector. 2017. Available online: https://www.eea.europa.eu/data-and-maps/indicators/energy-efficiency-and-energy-consumption/assessment-1 (accessed on 11 April 2019).
13. International Energy Outlook 2016 (IEO2016). 2016. Available online: https://www.eia.gov/outlooks/ieo/pdf/transportation.pdf (accessed on 28 April 2019).
14. Muneer, T.; Milligan, R.; Smith, I.; Doyle, A.; Pozuelo, M.; Knez, M. Energetic, environmental and economic performance of electric vehicles: Experimental evaluation. *Transp. Res. Part D Transp. Environ.* **2015**, *35*, 40–61. [CrossRef]
15. Akhavan-Rezay, E.; El-Saadany, E.F. Managing demand for plug-in electric vehicles in unbalanced LV systems with photovoltaics. *IEEE Trans. Ind. Inform.* **2017**, *13*, 1057–1067. [CrossRef]
16. Du, J.; Ouyang, D. Progress of Chinese electric vehicles industrialization in 2015: A review. *Appl. Energy* **2017**, *188*, 529–546. [CrossRef]
17. European Commission. White Paper 2011. 2018. Available online: https://ec.europa.eu/transport/themes/strategies/2011_white_paper_en (accessed on 20 May 2019).
18. Burchart-Korol, D.; Jursova, S.; Folęga, P.; Korol, J.; Pustejovska, P.; Blaut, A. Environmental life cycle assessment of electric vehicles in Poland and the Czech Republic. *J. Clean. Prod.* **2018**, *202*, 476–487. [CrossRef]
19. Chrisafis, A.; Vaughan, A. France to Ban Sales of Petrol and Diesel Cars by 2040. The Guardian. 2017. Available online: www.theguardian.com/business/2017/jul/06/france-ban-petrol-diesel-cars-2040-emmanuel-macron-volvo (accessed on 6 April 2019).
20. Braun, A.; Rid, W. Assessing driving pattern factors for the specific energy use of electric vehicles: A factor analysis approach from case study data of the Mitsubishi i–MiEV minicar. *Transp. Res. Part D Transp. Environ.* **2018**, *58*, 225–238. [CrossRef]
21. Petroff, A. These Countries Want to Ban Gas and Diesel Cars. CNN Money. 2017. Available online: http://money.cnn.com/2017/07/26/autos/countries-that-are-banning-gas-cars-for-electric/index.html (accessed on 18 March 2019).
22. Borén, S.; Nurhadi, L.; Ny, H.; Robèrt, K.; Broman, G.; Trygg, L. A strategic approach to sustainable transport system development—Part 2: The case of a vision for electric vehicle systems in southeast Sweden. *J. Clean. Prod.* **2017**, *140*, 62–71. [CrossRef]
23. Asthana, A.; Taylor, M. Britain to Ban Sale of all Diesel and Petrol Cars and Vans from 2040. The Guardian. 2017. Available online: http://www.theguardian.com/politics/2017/jul/25/britain-to-ban-sale-of-all-diesel-and-petrol-cars-and-vans-from-2040 (accessed on 6 April 2019).
24. Committee on Climate Change. Reducing Emissions in Scotland: 2015 Progress Report 2015. Available online: https://www.theccc.org.uk/wp-content/uploads/2015/01/Scotland-report-v6-WEB.pdf (accessed on 6 April 2019).
25. Scotland, A. *Reducing Scottish Greenhouse Gas Emissions*; Audit Scotland: Edinburgh, UK, 2011.
26. Transport Scotland. Scottish Household Survey. Travel Diary Results. 2016. Available online: https://www.transport.gov.scot/media/39692/sct09170037961.pdf (accessed on 14 May 2019).
27. Morrissey, P.; Weldon, P.; O'Mahony, M. Future standard and fast charging infrastructure planning: An analysis of electric vehicle charging behaviour. *Energy Policy* **2016**, *89*, 257–270. [CrossRef]

28. Koç, Ç.; Bektaş, T.; Jabali, O.; Laporte, G. The impact of depot location, fleet composition and routing on emissions in city logistics. *Transp. Res. Part B Methodol.* **2016**, *84*, 81–102. [CrossRef]
29. Goeke, D.; Schneider, M. Routing a mixed fleet of electric and conventional vehicles. *Eur. J. Oper. Res.* **2015**, *245*, 81–99. [CrossRef]
30. Pelletier, S.; Jabali, O.; Laporte, G.; Veneroni, M. Battery degradation and behaviour for electric vehicles: Review and numerical analyses of several models. *Transp. Res. Part B Methodol.* **2017**, *103*, 158–187. [CrossRef]
31. Pelletier, S.; Jabali, O.; Laporte, G. 50th Anniversary invited article—Goods distribution with electric vehicles: Review and research perspectives. *Transp. Sci.* **2016**, *50*, 3–22. [CrossRef]
32. Schiffer, M.; Walther, G. The electric location routing problem with time windows and partial recharging. *Eur. J. Oper. Res.* **2017**, *260*, 995–1013. [CrossRef]
33. Cubito, C.; Millo, F.; Boccardo, G.; Di Pierro, G.; Ciuffo, B.; Fontaras, G.; Serra, S.; Garcia, M.O.; Trentadue, G. Impact of different driving cycles and operating conditions on CO_2 emissions and energy management strategies of a euro-6 hybrid electric vehicle. *Energies* **2017**, *10*, 1590. [CrossRef]
34. Huertas, J.I.; Giraldo, M.; Quirama, L.F.; Díaz, J. Driving cycles based on fuel consumption. *Energies* **2018**, *11*, 3064. [CrossRef]
35. Zhang, Q.; Deng, W. An adaptive energy management system for electric vehicles based on driving cycle identification and wavelet transform. *Energies* **2016**, *9*, 341. [CrossRef]
36. Transport and Environment (T&E). *Mind the Gap! Why Official Car Fuel Economy Figures Don't Match up to Reality*; Dings, J., Ed.; Transport and Environment: Brussels, Belgium, 2013.
37. André, M. The ARTEMIS European driving cycles for measuring car pollutant emissions. *Sci. Total. Environ.* **2004**, *334*, 73–84. [CrossRef] [PubMed]
38. Nissan Leaf. 2018. Available online: https://www.electrocoches.eu/marcas/nissan/leaf/#nissan-leaf-24-kwh-acenta (accessed on 30 May 2019).
39. Renault Kangoo ZE. 2018. Available online: https://www.electrocoches.eu/marcas/renault/kangoo-ze/#renault-kangoo-z-e-kangoo-z-e (accessed on 30 May 2019).
40. Nissan e-NV200. 2018. Available online: https://www.electrocoches.eu/marcas/nissan/e-nv200/#nissan-e-nv200-combi-comfort-5-24-kwh (accessed on 30 May 2019).
41. Chaari, H.; Ballot, E. Fuel consumption assessment in delivery tours to develop eco driving behaviour. In Proceedings of the European Transport Conference 2012, Glasgow, Scotland, UK, 8–10 October 2012.
42. Andre, M. *Driving Cycle Development: Characterization of the Methods*; No. 961112; SAE Technical Paper: Warrendale, PA, USA, 1996.
43. Milligan, R. 13—Drive cycles for battery electric vehicles and their fleet management. In *Electric Vehicles: Prospects and Challenges*; Elsevier: Amsterdam, The Netherlands, 2017; pp. 489–555.
44. Esteves-Booth, A.; Muneer, T.; Kirby, H.; Kubie, J.; Hunter, J. The measurement of vehicular driving cycle within the city of Edinburgh. *Transp. Res. Part D Transp. Environ.* **2001**, *6*, 209–220. [CrossRef]
45. Alahmer, A.; Mayyas, A.; Mayyas, A.A.; Omar, M.; Shan, D. Vehicular thermal comfort models; A comprehensive review. *Appl. Therm. Eng.* **2011**, *31*, 995–1002. [CrossRef]
46. Nissan LEAF Owner's Manual. Available online: https://www.nissan.ca/content/dam/nissan/ca/owners/manuals/LEAF/2015-Nissan-LEAF.pdf (accessed on 30 May 2019).
47. Nissan UK. Nissan LEAF. 2016. Available online: https://www.nissan.co.uk/?&cid-psmM9WSXFpD_dc/D (accessed on 30 May 2019).
48. Rubin, E.; Davidson, C. *Introduction to Engineering and the Environment*; McGraw-Hill: New York, NY, USA, 2001.
49. Milligan, R. Critical Evaluation of the Battery Electric Vehicle for Sustainable Mobility. Ph.D. Thesis, Edinburgh Napier University, Edinburgh, Scotland, UK, 2016.
50. Wu, X.; Freese, D.; Cabrera, A.; Kitch, W.A. Electric vehicles' energy consumption measurement and estimation. *Transp. Res. Part D Transp. Environ.* **2015**, *34*, 52–67. [CrossRef]
51. Fiori, C.; Ahn, K.; Rakha, H.A. Power-based electric vehicle energy consumption model: Model development and validation. *Appl. Energy* **2016**, *168*, 257–268. [CrossRef]

52. Hayes, J.G.; Davis, K. Simplified electric vehicle powertrain model for range and energyconsumption based on EPA coastdown-parameters and test validation by Argonne National Lab data on the Nissan Leaf. In Proceedings of the IEEE Transportation Electrification Conference and Expo (ITEC), Dearborn, MI, USA, 15–18 June 2014; pp. 1–6. [CrossRef]
53. Ecotricity. Britain's Green Energy. 2016. Available online: https://www.ecotricity.co.uk/for-the-road/our-electric-highway (accessed on 15 May 2019).
54. Scottish Government. Scottish Greenhouse Gas Emissions Annual Target Report: 2015. 2016. Available online: https://beta.gov.scot/publications/scottish-greenhouse-gas-emissions-annual-target-report-2015/pages/6/ (accessed on 28 March 2019).
55. Gov. UK. Greenhouse Gas Reporting—Conversion Factors 2015. 2016. Available online: https://www.gov.uk/government/publications/greenhouse-gas-reporting-conversion-factors-2015 (accessed on 28 March 2019).
56. Electricity Distribution Losses—A Consultation Document. Available online: https://www.ofgem.gov.uk/ofgem-publications/44682/1362-03distlossespdf (accessed on 15 July 2019).
57. Lam, L.; Bauer, P.; Kelder, E. A practical circuit-based model for Li-ion battery cells in electric vehicle applications. In Proceedings of the IEEE 33rd International Telecommunications Energy Conference (INTELEC), Amsterdam, The Netherlands, 9–13 October 2011; pp. 1–9. [CrossRef]
58. Sauer, D.U. Batteries—Chargedischarge Curves. In *Encyclopedia of Electro-Chemical Power Sources*; Garche, J., Dyer, C., Moseley, P., Ogumi, Z., Rand, D., Scrosati, B., Eds.; Elsevier: Amsterdam, The Netherlands, 2009; pp. 443–451.
59. Bergveld, H.J.; Kruijt, W.S.; Notten, P.H.L. Battery management systems. *Battery Manag. Syst.* **2002**, *1*, 9–30.
60. Moura, S.J.; Fathy, H.K.; Callaway, D.S.; Stein, J.L. A stochastic optimal control approach for power management in plug-in hybrid electric vehicles. *IEEE Trans. Control Syst. Technol.* **2011**, *19*, 545–555. [CrossRef]
61. Campbell, R. Battery Characterization and Optimization for Use in Plug-In Hybrid Electric Vehicles: Hardware-in-the-Loop Duty Cycle Testing. Master's Thesis, Queen's University, Kingston, ON, Canada, 2011. Available online: https://pdfs.semanticscholar.org/b29c/65f2cdd44532022127466a291b0666f0870c.pdf?_ga=2.73880017.2138972187.1560500948-1227402376.1560500948 (accessed on 23 May 2019).
62. Khajepour, A.; Fallah, S.; Goodarzi, A. *Electric and Hybrid Vehicles—Technologies, Modeling, and Control: A Mechatronic Approach*; John Wiley & Sons: Hoboken, NJ, USA, 2014.
63. Neubauer, J.; Brooker, A.; Wood, E. Sensitivity of battery electric vehicle economics to drive patterns, vehicle range, and charge strategies. *J. Power Sources* **2012**, *209*, 269–277. [CrossRef]
64. Muneer, T.; Clarke, P.; Cullinane, K. The electric scooter as a means of green transport. In Proceedings of the IMechE conference on low carbon vehicles, London, UK, 20–21 May 2009.
65. Lane, B. Ford Focus Emissions. Next Green Car. 2016. Available online: http://www.nextgreencar.com/view-car/53716/ford-focus-1.6-style-105ps-s6-petrol-manual-5-speed (accessed on 30 May 2019).

© 2019 by the authors. Licensee MDPI, Basel, Switzerland. This article is an open access article distributed under the terms and conditions of the Creative Commons Attribution (CC BY) license (http://creativecommons.org/licenses/by/4.0/).

Article

A Variational Bayesian and Huber-Based Robust Square Root Cubature Kalman Filter for Lithium-Ion Battery State of Charge Estimation

Jing Hou [1,†], He He [2], Yan Yang [1,*], Tian Gao [1] and Yifan Zhang [1]

1. School of Electronic and Information, Northwestern Polytechnical University, Xi'an 710072, China; jhou0825@nwpu.edu.cn (J.H.); tiangao@nwpu.edu.cn (T.G.); shu@mail.nwpu.edu.cn (Y.Z.)
2. System Engineering Research Institute of CSSC, Beijing 100094, China; hehe5951@163.com
* Correspondence: yangyan7003@nwpu.edu.cn
† Current address: No.127, Youyi West Road, Xi'an 710072, China.

Received: 2 March 2019; Accepted: 30 April 2019; Published: 7 May 2019

Abstract: An accurate state of charge (SOC) estimation is vital for safe operation and efficient management of lithium-ion batteries. To improve the accuracy and robustness, an adaptive and robust square root cubature Kalman filter based on variational Bayesian approximation and Huber's M-estimation (VB-HASRCKF) is proposed. The variational Bayesian (VB) approximation is used to improve the adaptivity by simultaneously estimating the measurement noise covariance and the SOC, while Huber's M-estimation is employed to enhance the robustness with respect to the outliers in current and voltage measurements caused by adverse operating conditions. A constant-current discharge test and an urban dynamometer driving schedule (UDDS) test are performed to verify the effectiveness and superiority of the proposed algorithm by comparison with the square root cubature Kalman filter (SRCKF), the VB-based SRCKF, and the Huber-based SRCKF. The experimental results show that the proposed VB-HASRCKF algorithm outperforms the other three filters in terms of SOC estimation accuracy and robustness, with a little higher computation complexity.

Keywords: state of charge (SOC); lithium-ion battery; square root cubature Kalman filter (SRCKF); variational Bayesian approximation; Huber's M-estimation; adaptive; robust

1. Introduction

Lithium-ion batteries have been prevalently employed as energy storage devices in electric vehicles (EVs) and renewable energy storage systems owing to their high energy density, low self-discharge rate, and long cycle life [1]. The state of charge (SOC), which represents the amount of charge remaining in a battery, is one of the most important indicators of the current performance of the battery. Therefore, the SOC needs to be accurately estimated by a battery management system (BMS) in order to achieve battery equalization, charging/discharging control, and driving distance forecast for electric vehicles. In addition, for the battery energy storage system (BESS) of the grid, accurate estimation of SOC is crucial for controlling the SOC level within a certain range in order to reserve a certain amount of energy for load leveling as well as minimize battery health degradation. Hence, the battery charging process is only allowed during off-peak hours if the SOC level of the battery is below a specified maximum value. Meanwhile, the discharge is activated during the peak hours for load leveling and prevented if the SOC is lower than a specified minimum value.

Nevertheless, it is difficult to directly measure the SOC since the battery itself is a highly nonlinear and time-varying system on account of its complicated internal electrochemical reaction. Moreover, SOC estimation accuracy is affected by various factors such as ambient temperature, battery aging, and charging/discharging current rate[2]. Therefore, robust and accurate estimation of battery SOC is a difficult problem yet to be adequately resolved.

A great number of approaches have been proposed to estimate the SOC of lithium-ion batteries. The open-circuit voltage (OCV) method is a very simple one. But it needs a long rest time to measure the terminal voltage, making it almost impossible for moving vehicles. The coulomb counting (CC) method is widely applied in commercial BMSs, but its estimation accuracy depends heavily on the choice of the initial SOC values and suffers from measurement errors and accumulated errors. Machine learning algorithms including artificial neural networks (ANNs) [3–5], fuzzy logic (FL) algorithms [6–8], and support vector machines (SVMs) [9–11] have also been used for SOC estimation. These methods can estimate the SOC accurately and have no need for detailed physical knowledge of the battery. However, they require a large amount of experimental data to train the intelligent model beforehand. Moreover, the algorithms are easily divergent if the training data cannot completely cover the actual operating conditions [12].

Recently, model-based methods including observers [13,14], Kalman filter (KF), and its derivatives have been widely applied to SOC estimation. As the most representative method, the extended Kalman filter (EKF) was firstly introduced to estimate the SOC of a lithium-ion polymer battery by Plett [15] in 2004. But, EKF utilizes the first-order or second-order terms of Taylor's formula for linearization, which may bring in large linearization errors and thus degrade the SOC estimation accuracy. To overcome these defects, the unscented Kalman filter (UKF) [16–18] was put forward, which has a higher order of accuracy in estimating the mean and the error covariance of a state than EKF and does not need to calculate the Jacobian matrix. Ref. [19] evaluated the SOC estimation performance of the EKF, UKF, and particle filter (PF) for lithium-bismuth liquid metal batteries and showed that the UKF gave the most robust and accurate performance. Later, the cubature Kalman filter (CKF) [20,21] was proposed to improve the convergence rate and SOC estimation accuracy. CKF is based on the radial-spherical cubature rule, and is more suitable for state estimation of high-order nonlinear systems. Moreover, the cubature points and weights are uniquely determined by the dimension of the state, thus there is no need to tune parameters, as opposed to the UKF. Hence, it is easier for implementation. But asymmetric or non-positive definite covariance often arises during the iteration of CKF, which results in accuracy degradation or iteration interruption. To cope with the numerical instability of the CKF, square root cubature Kalman filter (SRCKF) [22,23] was developed, which can ensure the positive definiteness of the covariance. However, these KF-based methods also have some accuracy limitations. First, they are very sensitive to model mismatch, which can easily lead to filter divergence. Second, the process and measurement noise statistics are assumed to be known and Gaussian, which is not necessarily true in practical applications.

As we know that battery parameters change with SOC, temperature, and battery aging during the charge and discharge process, it is not easy to establish an exactly matched battery model. In other words, modeling error is usually inevitable. In addition, in complex industrial applications, the precision of current and voltage sensors is much lower than in the laboratory. Moreover, sensor noise is probably a non-Gaussian process with unknown or time-varying covariance. Therefore, to account for modeling error and contaminated distributions or outliers in measurements, an adaptive and robust SOC estimation method is badly required. To address the adaptivity, an adaptive square root unscented Kalman filter (ASRUKF) [24] and an adaptive cubature Kalman filter (ACKF) [21] based on the improved Sage–Husa estimator were presented. These methods adaptively adjusted the values of the process and measurement covariances in the estimation process to improve the accuracy of SOC estimation. El Din et al. [25] proposed a multiple-model EKF (MM-EKF) and an autocovariance least squares (ALS) method for estimating the SOC under measurement noise statistic uncertainties. MM-EKF reduced the impact caused by unknown measurement noise statistics by calculating the weighted sum of the estimates of multiple hypothesized EKFs. The ALS method extracted the possible correlation in the innovation sequence to estimate the measurement noise covariance. On the other hand, in order to improve the robustness of SOC estimation, H infinity filters [1,26–28] were employed to deal with gross errors or outliers in measurements. However, when there are both modeling errors and outliers in the battery system, those filters cannot achieve satisfactory accuracy.

Actually, the variational Bayesian (VB)-based filter [29–33] is one of the most general adaptive filters. Its adaptive strategy has a strong ability to track the time-varying measurement noise covariance. Meanwhile, Huber's M-estimator [34] is a combined minimum l_1 and l_2 norm estimation technique, which exhibits robustness with respect to the contaminated distributions and outliers by modifying the filtering update. Thus, it will be promising to combine VB approximation and Huber's M-estimation to achieve both adaptivity and robustness. Li et al. [30] has acted out this idea in the framework of UKF. The efficiency of the proposed filter was verified through the numerical simulation test. In this paper, we extend this idea within the framework of SRCKF for adaptive and robust SOC estimation of the lithium-ion batteries. The measurement noise covariance is simultaneously estimated with the SOC to account for battery model uncertainties and measurement noise covariance uncertainties. Meanwhile, the outliers in current and voltage measurements caused by adverse operating conditions are accounted for by Huber's M-estimation. The effectiveness of the proposed filter has been verified through experiments under different operating conditions. The results show that the proposed filter can achieve much better estimation accuracy than SRCKF, the VB-based SRCKF, and the Huber-based SRCKF with a little higher computation complexity when there are both measurement outliers and mistuned measurement noise covariance.

The contributions of this paper are the following: (1) An adaptive and robust square root cubature Kalman filter based on variational Bayesian approximation and Huber's M-estimation (VB-HASRCKF) is proposed for SOC estimation of lithium-ion batteries; (2) Compared with the existing models, the case that there are both modeling errors and outliers in SOC estimation is firstly taken into account and the VB-HASRCKF is used to handle this problem; (3) A constant-current discharge test and an UDDS test are performed to verify the effectiveness and superiority of the proposed algorithm by comparison with the square root cubature Kalman filter (SRCKF), the VB-based SRCKF, and the Huber-based SRCKF.

The organization of this paper proceeds as follows: Section 2 describes the battery model and the parameter identification. Section 3 illustrates the VB-based adaptive SRCKF algorithm. In Section 4, the adaptive and robust SRCKF based on VB approximation and Huber's M-estimation is presented. The experimental verification and analysis are presented in Section 5. Finally, Section 6 provides a conclusion.

2. Battery Modeling and Parameter Identification

2.1. Battery Modeling

For the accurate estimation of the SOC, a reliable battery model is required. The existing models include the electrochemical model [35] and the equivalent circuit model (ECM) [24,28]. Among others, the ECMs have a better trade-off between accuracy and complexity. Therefore, we adopted a typical ECM, the first order resistor–capacitor (RC) model as shown in Figure 1, to model the lithium-ion battery in this paper.

The electrical behavior of the model can be written as follows:

$$U_t = U_{oc} - U_1 - I_L R_0,$$
$$\dot{U}_1 = \frac{I_L}{C_1} - \frac{U_1}{R_1 C_1},$$

where, U_t denotes the terminal voltage of the battery, U_{oc} is the open-circuit voltage, U_1 is the polarization voltage of the RC network, I_L is the load current, R_0 represents the ohmic internal resistance, and R_1 and C_1 represent the polarization resistance and polarization capacitance, respectively.

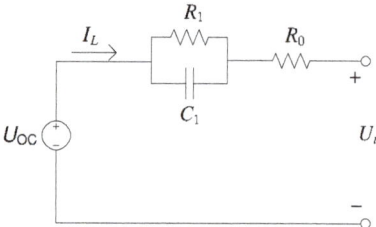

Figure 1. A first-order resistor–capacitor (RC) battery model.

The state of charge (SOC) is defined as the ratio of the remaining capacity in a battery over the maximum available capacity. Using the CC method, the battery SOC can be calculated as

$$SOC_t = SOC_{t_0} - \frac{1}{Q_{max}} \int_{t_0}^{t} \eta_c I_{L,t} dt,$$

where $I_{L,t}$ is the load current at time t, Q_{max} is the maximum available capacity of the battery, and η_c is the coulomb efficiency (CE), which is defined as the ratio of the discharge capacity to the charge capacity of the battery in the same cycle.

Here, we offer some explanations about CE. Theoretically, when a cell is free of undesired side reactions, its CE should permanently be 1.00 [36]. In fact, the CE value is not always equal to 1, and it is related to many battery parameters such as SOC, current rate, operating temperature, and battery capacity degradation. Zheng et al. [37] showed the correlation between SOC and CE and claimed that CE is almost invariant with SOC changes. The correlations of CE with current rate and operating temperature were discussed in [38,39]. Yang et al. [36] gave in-depth discussions regarding battery degradation, aging mechanisms, and CE evolution, and observed that a decrease of the CE value corresponds to an increase in the degradation rate and that an increase of the CE value indicates a decrease in the degradation rate. Meanwhile, experimental results also suggest that CE evolution is distinct for different types of batteries. Here, we do not consider the impact of these factors on CE and assume $\eta_c = 1$ since the CE of the lithium-ion battery is usually very high and has a fairly small measurement error. Moreover, this paper mainly focuses on handling the model uncertainties, which might also include the uncertainty brought by an inaccurate CE.

Taking $x = [SOC, U_1]^T$ as the state vector, the load current I_L as the input, and the terminal voltage U_t as the output, we can obtain the discrete state-space model as

$$x_{k+1} = f(x_k, I_{L,k}) + w_k,$$
$$z_k = U_{t,k} = h(x_k, I_{L,k}) + v_k,$$

where $w_k \sim \mathcal{N}(0, Q_k)$ is the Gaussian process noise with covariance Q_k, and $v_k \sim \mathcal{N}(0, R_k)$ is the measurement noise with variance R_k. $f(\cdot)$ and $h(\cdot)$ represent the nonlinear functions of state vector x_k and input $I_{L,k}$. Their mathematical expressions are

$$f(\cdot) = \begin{bmatrix} 1 & 0 \\ 0 & e^{-\frac{\Delta t}{\tau_1}} \end{bmatrix} \begin{bmatrix} SOC_k \\ U_{1,k} \end{bmatrix} + \begin{bmatrix} -\frac{\eta \Delta t}{Q_{max}} & 0 \\ 0 & R_1(1 - e^{-\frac{\Delta t}{\tau_1}}) \end{bmatrix} I_{L,k},$$

$$h(\cdot) = U_{OC}(SOC_k) - U_{1,k} - I_{L,k} R_0,$$

where Δt is the sampling interval of the current, and $\tau_1 = R_1 C_1$ is the time constant of the RC network. $U_{OC}(SOC_k)$ represents the nonlinear relationship between the open-circuit voltage and the SOC, which is determined by experiments in the next section.

2.2. OCV–SOC Relationship Determination

In order to acquire the relationship between OCV and SOC, the following test procedure was designed:

(1) First, fully charge the battery and rest for 1 h to finish the process of depolarization. Then the measured terminal voltage is assumed to be the discharge OCV value.
(2) Discharge the battery at a constant current of 1 A until 10% of the maximum available capacity is consumed, and measure the OCV after resting for 1 h.
(3) Repeat step (2) until the battery reaches its lower cut-off voltage.

Based on the measured data, a fifth-order polynomial in Equation (1) is selected to characterize the relationship between OCV and SOC, and the measured data and fitted curve are presented in Figure 2. The R-squared is used to represent the goodness of fit. The normal value range of the R-squarde is $0 - 1$ and the closer to 1, the better for curve fitting [40]. It can be seen that the curve fits well with the measurement data, indicating that the selected fifth-order polynomial model can describe the relationship between OCV and SOC very well.

$$U_{oc}(SOC) = 3.083 + 4.859 \times SOC - 18.21 \times SOC^2 + 38.56 \times SOC^3 \\ - 38.64 \times SOC^4 + 14.58 \times SOC^5. \tag{1}$$

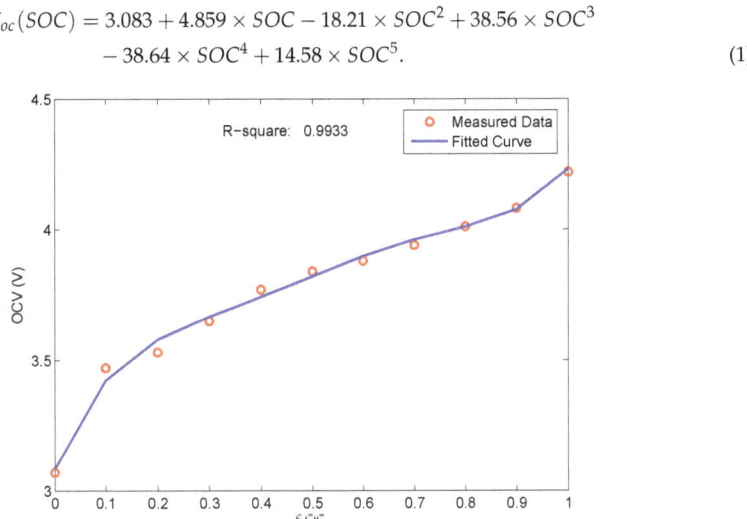

Figure 2. The relationship curve of open-circuit voltage (OCV) versus state of charge (SOC).

2.3. Online Parameter Identification

The battery model parameters are changeable with the operating current, SOC, temperature and aging, so that offline identification of model parameters will inevitably lead to some model errors. Therefore, in this paper the model parameters are updated online using the forgetting factor recursive least squares (FFRLS) algorithm [21] to improve the SOC estimation accuracy.

The transfer function of the battery model can be written as

$$G(s) = \frac{(U_{oc} - U_t)(s)}{I_L(s)} = \frac{R_1}{1 + R_1 C_1 s} + R_0. \tag{2}$$

Using $X(s) = \frac{X(k) - X(k-1)}{T}$ to discretize Equation (2), where T is the sampling period, we can obtain

$$(\frac{R_1 C_1}{T} + 1)(U_{oc} - U_t)(k) = \frac{R_1 C_1}{T}(U_{oc} - U_t)(k-1) + (\frac{R_0 R_1 C_1}{T} + R_0 + R_1)I_L(k) - \frac{R_0 R_1 C_1}{T} I_L(k-1). \tag{3}$$

Replacing the coefficients in Equation (3) with k_1, k_2, and k_3, we can get

$$(U_{oc} - U_t)(k) = -k_1(U_{oc} - U_t)(k-1) + k_2 I_L(k) + k_3 I_L(k-1). \tag{4}$$

Define

$$\begin{cases} \varphi(k) = [-(U_{oc} - U_t)(k-1); I_L(k); I_L(k-1)] \\ \theta(k) = [k_1; k_2; k_3] \\ y(k) = (U_{oc} - U_t)(k). \end{cases}$$

The discrete time-domain difference expression in Equation (4) can be rewritten as

$$y(k) = \varphi^T(k)\theta(k) + \zeta(k), \tag{5}$$

where ζ is a Gaussian random noise with zero mean.

The vector $\theta(k)$ in Equation (5) can be solved using the FFRLS algorithm with forgetting factor λ (typically $\lambda = [0.95, 1]$), formulated as

$$\begin{cases} \hat{\theta}(k) = \hat{\theta}(k-1) + K(k)[y(k) - \varphi^T(k)\hat{\theta}(k-1)] \\ K(k) = \frac{P(k-1)\varphi(k)}{\lambda + \varphi^T(k)P(k-1)\varphi(k)} \\ P(k) = \frac{1}{\lambda}[I - K(k)\varphi^T(k)]P(k-1). \end{cases}$$

Finally, R_0, R_1, and C_1 can be derived by

$$\begin{cases} R_0 = \frac{k_3}{k_1} \\ R_1 = \frac{k_2 - R_0}{k_1 + 1} \\ C_1 = \frac{-k_1 T}{(k_1 + 1)R_1}. \end{cases}$$

2.4. Model Validation

The battery model described in this paper was verified through a 1A constant-current discharge test and UDDS test. The measured terminal voltage and the model terminal voltage were compared, as shown in Figures 3 and 4. The maximum absolute error was 0.045 V, while the mean absolute error was 0.0017 V in the constant-current discharge test. The maximum and mean absolute errors were 0.091 V and 0.0047 V in the UDDS test, respectively. It is clear that the model terminal voltage agrees well with the measured voltage. This illustrates the effectiveness of the battery model.

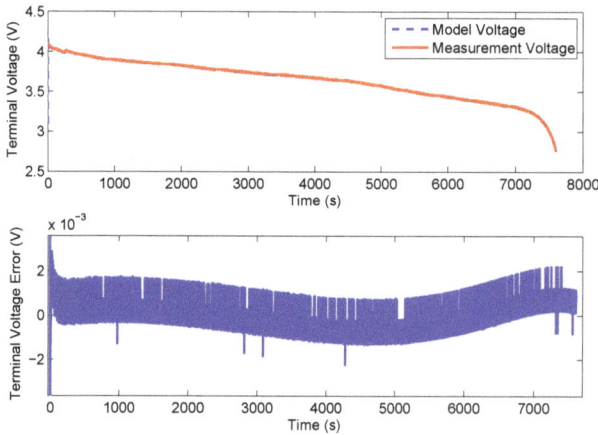

Figure 3. Experimental terminal voltage results in a constant-current discharge test.

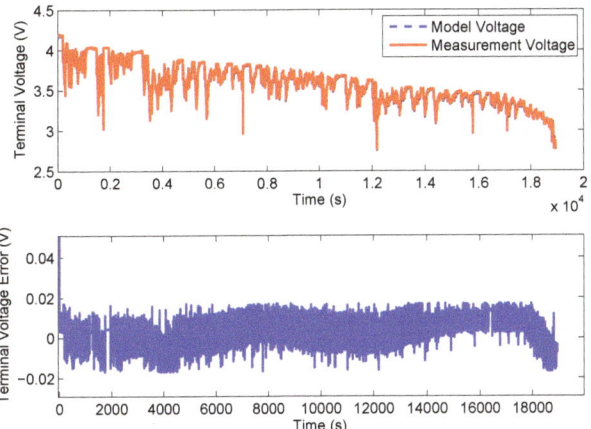

Figure 4. Experimental terminal voltage results in an urban dynamometer driving schedule (UDDS) test.

3. Variational Bayesian-Based Adaptive Square Root Cubature Kalman Filter

Recently, the variational Bayesian (VB)-based filter, which has a strong ability to track the uncertain parameters, has drawn extensive attention. In [29], a VB approximation-based adaptive Kalman filter was first presented, which can be used for the measurement of noise covariance adaptation. For nonlinear systems, a VB-based UKF is developed in [30]. Here, we focus on the adaptive square root cubature Kalman filter (SRCKF) based on VB approximation.

Let us rewrite the state-space equations for SOC estimation as

$$\begin{cases} x_{k+1} = f(x_k, I_{L,k}) + w_k \\ z_k = h(x_k, I_{L,k}) + v_k \\ w_k \sim \mathcal{N}(0, Q_k) \\ v_k \sim \mathcal{N}(0, R_k), \end{cases}$$

where the process noise covariance Q_k is assumed to be known, and the measurement noise covariance R_k is unknown.

The VB-based adaptive filter uses VB approximation to estimate the joint posterior distribution of the state and covariance $p(x_k, R_k | z_{1:k})$, as follows:

$$p(x_k, R_k | z_{1:k}) \approx Q_x(x_k) Q_R(R_k).$$

The VB approximation can now be formed by minimizing the Kullback–Leibler (KL) divergence between the separable approximation and the true posterior:

$$KL\left[Q_x(x_k)Q_R(R_k) || p(x_k, R_k | z_{1:k})\right]$$
$$= \int Q_x(x_k) Q_R(R_k) \times \log \left(\frac{Q_x(x_k) Q_R(R_k)}{p(x_k, R_k | z_{1:k})} \right) dx_k dR_k.$$

Minimizing the KL divergence with respect to the probability densities $Q_x(x_k)$ and $Q_R(R_k)$ in turn, while keeping the other fixed, we can get the following equations:

$$Q_x(x_k) \propto \exp \left(\int \log p(z_k, x_k, R_k | z_{1:k-1}) Q_R(R_k) dR_k \right),$$

$$Q_R(R_k) \propto \exp\left(\int \log p(z_k, x_k, R_k | z_{1:k-1}) Q_x(x_k) dx_k\right).$$

Computing the above equations, we can get the following densities [29]:

$$Q_x(x_k) = \mathcal{N}(x_k | \hat{x}_k, P_k),$$

$$Q_R(R_k) = \text{IW}(R_k | v_k, V_k),$$

where $\text{IW}(\cdot)$ represents the inverse Wishart (IW) distribution, and the parameters $\hat{x}_k$, P_k, v_k, and V_k can be calculated using Kalman-type filters.

The above are the basic idea of the adaptive Kalman filter based on VB approximation. To solve the nonlinear battery SOC estimation problem, the VB method is rewritten in the SRCKF framework in this paper. The filtering procedure of the VB-based adaptive SRCKF algorithm (VB-ASRCKF) is summarized as follows.

Step 1. Initialize state and parameter estimation: $\hat{x}_0$, S_0, Q_0, v_0, V_0.
Step 2. Predict ($k = 1, 2, 3, ...$).

Step 2.1 Calculate the cubature points

$$X_{i,k-1} = S_{k-1}\xi_i + \hat{x}_{k-1} \quad i = 1, 2, ...; 2n, \tag{6}$$

where ξ_i is the ith column of the weight matrix of the cubature points $[nI_n \ -nI_n]$, I_n is the $n \times n$ identity matrix, and n is the dimension of the state vector.

Step 2.2 Propagate the cubature points through the process equation and calculate the predicted state values:

$$X_{i,k|k-1} = f(X_{i,k-1}, I_{L,k-1}), \tag{7}$$

$$\hat{x}_{k|k-1} = \frac{1}{2n}\sum_{i=1}^{2n} X_{i,k|k-1}. \tag{8}$$

Step 2.3 Calculate the square root of the covariance of the predicted state

$$S_{k|k-1} = \text{Tria}([X^*_{k|k-1} \ S_{Q,k-1}]^T), \tag{9}$$

where $S = \text{Tria}(A)$ represents the QR decomposition of matrix A, which obtains a unitary matrix B and an upper triangular matrix C. We define $S = C^T$. $S_{Q,k-1}$ is the Cholesky decomposition of Q_{k-1}. That is, $Q_{k-1} = S_{Q,k-1}S_{Q,k-1}^T$. In addition,

$$X^*_{k|k-1} = \frac{1}{\sqrt{2n}}[X_{1,k|k-1} - \hat{x}_{k|k-1}, \ldots, X_{2n,k|k-1} - \hat{x}_{k|k-1}]. \tag{10}$$

Step 2.4 Calculate the parameters of the IW distribution of measurement noise covariance

$$v_{k|k-1} = \rho(v_{k-1} - n - 1) + n + 1, \tag{11}$$

$$V_{k|k-1} = BV_{k-1}B^T, \tag{12}$$

where ρ is a scale factor that $0 < \rho \leq 1$ and B is a matrix that $0 < |B| \leq 1$ with a reasonable choice for the matrix $B = \sqrt{\rho}I_d$. I_d is an identity matrix and d is the dimension of the measurement.

Step 3. Update: the update of VB-ASRCKF utilizes an iterate filtering framework.

Step 3.1 First set $\hat{x}_k^{(0)} = \hat{x}_{k|k-1}$, $S_k^{(0)} = S_{k|k-1}$, $V_k^{(0)} = V_{k|k-1}$, and $v_k = 1 + v_{k|k-1}$.

Step 3.2 Calculate the cubature points of the predicted state

$$X_{i,k|k-1} = S_{k|k-1}\xi_i + \hat{x}_{k|k-1}. \tag{13}$$

Step 3.3 Propagate the cubature points through the measurement equation and calculate the predicted measurement value

$$Z_{i,k|k-1} = h(X_{i,k|k-1}), \tag{14}$$

$$\hat{z}_{k|k-1} = \frac{1}{2n}\sum_{i=1}^{2n} Z_{i,k|k-1}. \tag{15}$$

Step 3.4 Calculate the covariance of the state and the measurement

$$P_{xz,k|k-1} = \chi_{k|k-1} Z_{k|k-1}^T, \tag{16}$$

where

$$\chi_{k|k-1} = \frac{1}{\sqrt{2n}}[X_{1,k|k-1} - \hat{x}_{k|k-1}, \ldots, X_{2n,k|k-1} - \hat{x}_{k|k-1}], \tag{17}$$

$$Z_{k|k-1} = \frac{1}{\sqrt{2n}}[Z_{1,k|k-1} - \hat{z}_{k|k-1}, \ldots, Z_{2n,k|k-1} - \hat{z}_{k|k-1}]. \tag{18}$$

Step 3.5 For $j = 1 : N$, iterate the following N (N denotes iterated times) steps.

Step 3.5.1 Calculate the measurement covariance

$$R_k^{(j)} = (v_k - n - 1)^{-1} V_k^{(j-1)}. \tag{19}$$

Step 3.5.2 Calculate the square root of the innovation covariance

$$S_{zz,k|k-1}^{(j)} = Tria([Z_{k|k-1} \ S_{R,k}^{(j)}]^T), \tag{20}$$

where $S_{R,k}^{(j)}$ is the Cholesky decomposition of $R_k^{(j)}$.

Step 3.5.3 Calculate the filter gain

$$K_k^{(j)} = \frac{P_{xz,k|k-1}/[S_{zz,k|k-1}^{(j)}]^T}{S_{zz,k|k-1}^{(j)}}. \tag{21}$$

Step 3.5.4 Calculate the state estimate and the square root of its covariance

$$\hat{x}_k^{(j)} = \hat{x}_{k|k-1} + K_k^{(j)}(z_k - \hat{z}_{k|k-1}), \tag{22}$$

$$S_k^{(j)} = Tria([\chi_{k|k-1} - K_k^{(j)} Z_{k|k-1} \ K_k^{(j)} S_{R,k}^{(j)}]^T). \tag{23}$$

Step 3.5.5 Calculate the updated parameter of the IW distribution of measurement noise covariance

$$X_{i,k}^{(j)} = S_k^{(j)} \xi_i + \hat{x}_k^{(j)}, \tag{24}$$

$$V_k^{(j)} = V_{k|k-1} + \frac{1}{2n}\sum_{i=1}^{2n}(z_k - h(X_{i,k}^{(j)}))(z_k - h(X_{i,k}^{(j)}))^T. \tag{25}$$

Step 3.6 Until $j = N$, set $\hat{x}_k = \hat{x}_k^{(N)}$, $S_k = S_k^{(N)}$, $V_k = V_k^{(N)}$, and end for. Then one cycle of the VB-ASRCKF algorithm is finished.

4. Variational Bayesian and Huber-Based Robust Square Root Cubature Kalman Filter

In adverse environments, the current and voltage sensor noises maybe non-Gaussian, or there may be outliers in the voltage and current measurements. In this case, the above adaptive filter is incapable of achieving good performance. As we know, Huber's M-estimation method is a combined minimum l_1 and l_2 norm estimation technology, which exhibits robustness with respect to non-Gaussian distributions and outliers [34]. Therefore, to account for the measurement noise covariance uncertainties as well as the non-Gaussian distributions or outliers in the measurements for the battery SOC estimation, VB approximation and Huber's M-estimation-based square root cubature Kalman filter (VB-HASRCKF) with adaptivity and robustness is proposed.

First, let's give an introduction to Huber's M-estimation. The update of the filter can be viewed as a solution to a particular weighted least squares problem; we can construct the nonlinear regression measurement model as

$$\begin{bmatrix} z_k \\ \hat{x}_{k|k-1} \end{bmatrix} = \begin{bmatrix} h(x_k) \\ x_k \end{bmatrix} + \begin{bmatrix} v_k \\ \delta \hat{x}_{k|k-1} \end{bmatrix}, \tag{26}$$

where $\hat{x}_{k|k-1}$ is the prediction of the state with covariance $P_{k|k-1}$, and $\delta \hat{x}_{k|k-1}$ is the error between the true state and its prediction. Through the definition of the quantities

$$W_k = \begin{bmatrix} R_k & 0 \\ 0 & P_{k|k-1} \end{bmatrix}, \tag{27}$$

$$y_k = W_k^{-1/2} \begin{bmatrix} z_k \\ \hat{x}_{k|k-1} \end{bmatrix}, \tag{28}$$

$$g(x_k) = W_k^{-1/2} \begin{bmatrix} h(x_k) \\ x_k \end{bmatrix}, \tag{29}$$

$$\varepsilon_k = W_k^{-1/2} \begin{bmatrix} v_k \\ \delta \hat{x}_{k|k-1} \end{bmatrix}, \tag{30}$$

the above Equation (26) is integrated into

$$y_k = g(x_k) + \varepsilon_k. \tag{31}$$

Define Huber's generalized cost function as

$$J(x_k) = \sum_{i=1}^{n+d} \rho(e_{i,k}), \tag{32}$$

where n and d are the dimensions of the state and the measurement, respectively. $e_k = y_k - g(x_k)$ is the residual error, $e_{i,k}$ is the ith component of e_k, and ρ is defined as

$$\rho(e_{i,k}) = \begin{cases} e_{i,k}^2/2 & |e_{i,k}| < \gamma \\ \gamma|e_{i,k}| - \gamma^2/2 & |e_{i,k}| \geq \gamma \end{cases}, \tag{33}$$

where γ is a tuning parameter to be chosen to give the desired efficiency to the Gaussian model.

Minimizing (32), we can find the implicit solution to the nonlinear regression problem as

$$\sum_{i=1}^{n+d} \phi(e_{i,k}) \frac{\partial e_{i,k}}{\partial x_i} = 0, \tag{34}$$

where $\phi(e_{i,k}) = \rho'(e_{i,k})$.

Define

$$\psi(e_{i,k}) = \phi(e_{i,k})/e_{i,k} = \begin{cases} 1 & |e_{i,k}| < \gamma \\ \text{sgn}(e_{i,k})\gamma/e_{i,k} & |e_{i,k}| \geq \gamma \end{cases}, \tag{35}$$

and let $\Psi = \text{diag}[\psi(e_{i,k})]$. Then, Ψ can be used to reformulate the measurement information. There are two ways to reformulate the measurement information.

One way is to re-weight the residual error covariance by assigning smaller weights to outlying observations. Denote $\tilde{W}_k$ as the modified measurement noise covariance, that is

$$\tilde{W}_k = W_k^{1/2} \Psi^{-1} (W_k^{1/2})^T.$$

Then, denote $\tilde{R}_k$ as the modified measurement covariance with $\tilde{R}_k = \tilde{W}_k(1:d, 1:d)$. Replacing R_k with $\tilde{R}_k$ will lead to the robust filter.

Another way is to re-construct the "pseudo observations" by truncating the too-large or too-small observations. When $|e_{i,k}| \geq \gamma$, replace $e_{i,k}$ with $\text{sign}(e_{i,k})\gamma$. Denote $\tilde{e}_k$ as the modified residual error and $\tilde{e}_k = \Psi e_k$, which is the same to modify the measurements y_k. Denote $\tilde{y}_k$ as the modified y_k by

$$\tilde{y}_k = g(x_k) + \tilde{e}_k. \tag{36}$$

We can get the modified original measurement $\tilde{z}_k$ as

$$\tilde{z}_k = W_k^{1/2} \tilde{y}_k(1:d) = W_k^{1/2} [g(x_k) + \Psi(1:d, 1:d) e_k(1:d)]. \tag{37}$$

Replacing z_k with $\tilde{z}_k$ will also lead to the robust filter. Moreover, the two methods have the same robust performance. In this paper, we will select the second method to modify the measurements. The detailed reasons can be found in [30].

Considering the case under which there is both unknown measurement noise covariance and outliers in the measurements, Huber's M-estimation is integrated with VB approximation within the SRCKF framework in this paper. The filtering procedure of the proposed VB HASRCKF algorithm is summarized as follows.

Step 1. Initialize: $\hat{x}_0, S_0, Q_0, v_0, V_0$.
Step 2. Predict: using equations (6)–(12) to obtain $\hat{x}_{k|k-1}, S_{k|k-1}, V_{k|k-1}$, and $v_{k|k-1}$.
Step 3. Update.

Step 3.1 Using (13)–(18) to obtain $Z_{k|k-1}, X_{k|k-1}, \hat{z}_{k|k-1}$, and $P_{xz,k|k-1}$.
Step 3.2 Set $\hat{x}_k^{(0)} = \hat{x}_{k|k-1}, S_k^{(0)} = S_{k|k-1}, V_k^{(0)} = V_{k|k-1}$, and $v_k = 1 + v_{k|k-1}$
Step 3.3 For $j = 1 : N$, iterate the following N (N denotes iterated times) steps.

Step 3.3.1 Use (19)–(21) to calculate $R_k^{(j)}, S_{zz,k|k-1}^{(j)}$ and $K_k^{(j)}$.
Step 3.3.2 Calculate $\tilde{z}_k$ using (26)–(35) and (36)–(37).

Step 3.3.3 Replace z_k with $\tilde{z}_k$ in (22)–(25). That is, proceed with the following calculations:

$$\hat{x}_k^{(j)} = \hat{x}_{k|k-1} + K_k^{(j)}(\tilde{z}_k - \hat{z}_{k|k-1}),$$
$$S_k^{(j)} = Tria([\chi_{k|k-1} - K_k^{(j)} Z_{k|k-1} \quad K_k^{(j)} S_{R,k}^{(j)}]^T),$$
$$X_{i,k}^{(j)} = S_k^{(j)} \xi_i + \hat{x}_k^{(j)},$$
$$V_k^{(j)} = V_{k|k-1} + \frac{1}{2n} \sum_{i=1}^{2n} (\tilde{z}_k - h(X_{i,k}^{(j)}))(\tilde{z}_k - h(X_{i,k}^{(j)}))^T,$$

where $S_{R,k}^{(j)}$ is the Cholesky decomposition of $R_k^{(j)}$.

Step 3.4 Until $j = N$, set $\hat{x}_k = \hat{x}_k^{(N)}$, $S_k = S_k^{(N)}$, $V_k = V_k^{(N)}$ and end for. Then one cycle of the VB-HASRCKF algorithm is finished.

5. Experimental Verification and Analysis

5.1. Experimental Settings

To evaluate the effectiveness of the proposed method, a test bench shown in Figure 5 is established. The test bench consists of a lithium-ion battery cell, an electronic load, and a host computer. The tested lithium-ion battery cell is type $ICR18650$ developed by SAMSUNG, whose nominal capacity is 2.6 Ah, nominal voltage is 3.63 V, and charging and discharging cutoff voltages are 4.2 V and 2.75 V, respectively. The type of electronic load is $IT8516S$ produced by ITECH, whose current measurement accuracy is $\pm(0.1\% + 0.1\%$ full scale) and voltage measurement accuracy is $\pm(0.02\% + 0.02\%$ full scale).

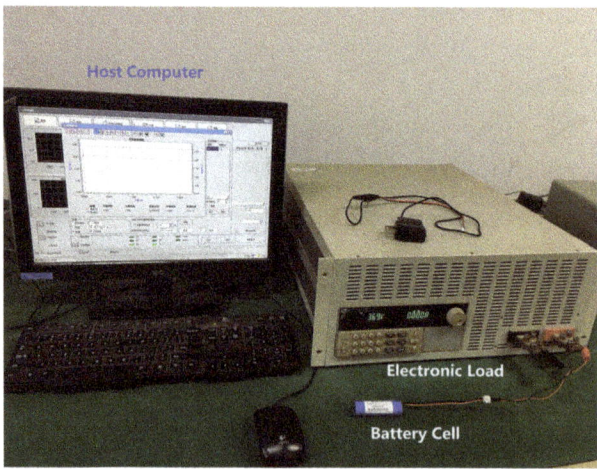

Figure 5. The experimental setup.

A constant-current discharge test and an urban dynamometer driving schedule (UDDS) test were performed to validate the proposed VB-HASRCKF algorithm on the basis of the test bench. The current, voltage, and SOC can be recorded via the host computer. The battery current and voltage were both sampled at 1 second. In each test, the true SOC was obtained using the CC method, and the estimation accuracy and robustness of the proposed VB-HASRCKF were evaluated by comparison with SRCKF, Huber-based SRCKF (HSRCKF), and VB-ASRCKF under different tests and different kinds of violations:

Case a: without any outliers or mistuning;
Case b: with mistuned measurement noise covariance;
Case c: with outliers in the measurements;
Case d: with both mistuned measurement noise covariance and outliers in the measurements.

5.2. SOC Estimation Experimental Results under a 1A Constant-Current Discharge Test

The experiment was performed with a constant discharge current of 1A. The initial SOC value was set to be 0.8, different from the real SOC of 1.0. The process noise covariance was set as $Q_k = 1 \times 10^{-8} I_2$. The measurement noise variance used for SRCKF and HSRCKF was assumed to be $R_k = 0.01$ in cases a and c, while mistuned to $R_k = 0.1$ in cases b and d. The initial parameter values of VB-based filters were $v_0 = 100$ and $V_0 = 1$ in cases a and c, and $v_0 = 100$ and $V_0 = 10$ in cases b and d. For cases c and d, some outliers in the voltage and current measurements were artificially added on the basis of the real experiment data. Figure 6 presents the contaminated current and voltage measurements. The outlying voltage measurements began at time k = 100 s, 1000 s, 2000 s, and 3000 s, and lasted for 3~10 s. The outliers in the current measurements were added at time k = 100 s and k = 1000 s.

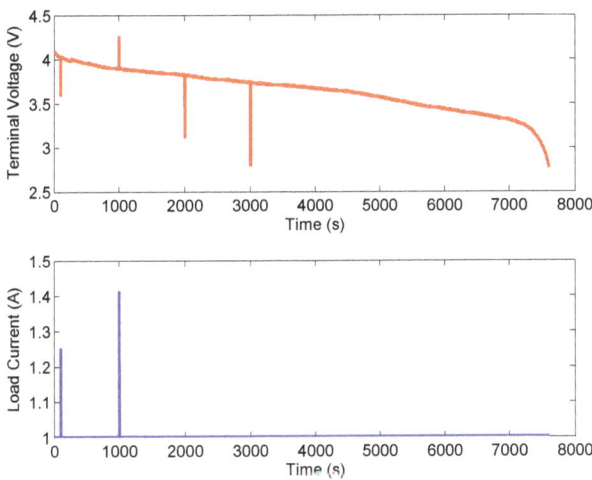

Figure 6. The outlying voltage and current measurements in a constant-current discharge test under cases c and d.

The SOC estimation results using SRCKF, HSRCKF, VB-ASRCKF, and VB-HASRCKF under four cases are shown in Figures 7 and 8. The maximum and mean absolute estimation errors are shown in Table 1. Meanwhile, the estimated measurement variances are presented in Figure 9. As can be seen in Figure 7a and Table 1, in case a, the mean absolute error of the four filters is around 0.70%, so they have comparable SOC estimation accuracy. Meanwhile, the converge rates of all the filters are also comparable since the SOC estimation errors are almost immediately decreased to under 5% for all filters. In addition, as shown in Figure 9, the estimated measurement noise variances of VB-ASRCKF and VB-HASRCKF are about 0.02, which does not deviate too much from the assumed variances of SRCKF and HSRCKF. It shows that the assumed measurement noise variance is consistent with the actual value. Therefore, the four filters display comparable performance.

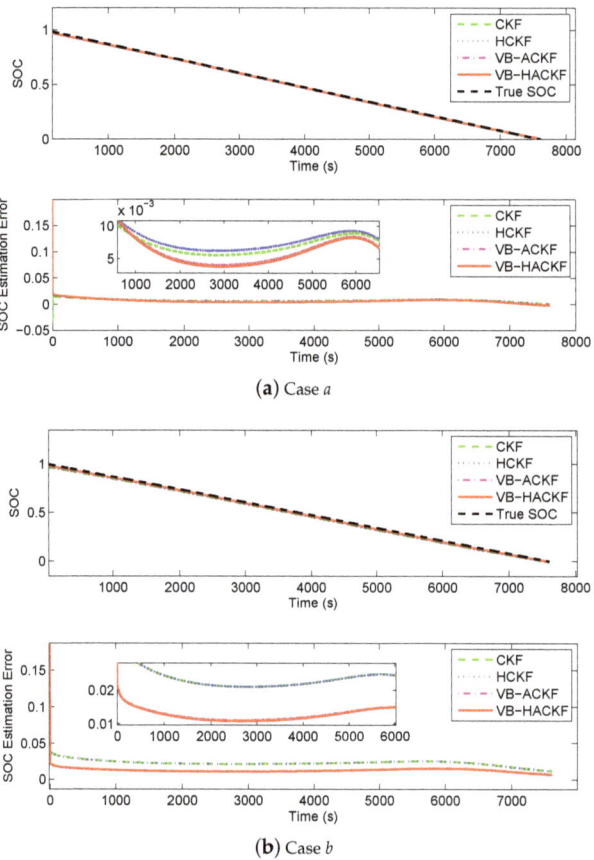

Figure 7. SOC estimation results of four filters in a constant-current discharge test under cases *a* and *b*.

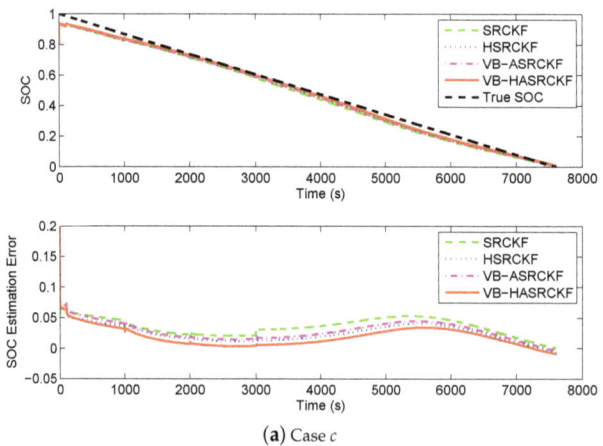

Figure 8. *Cont.*

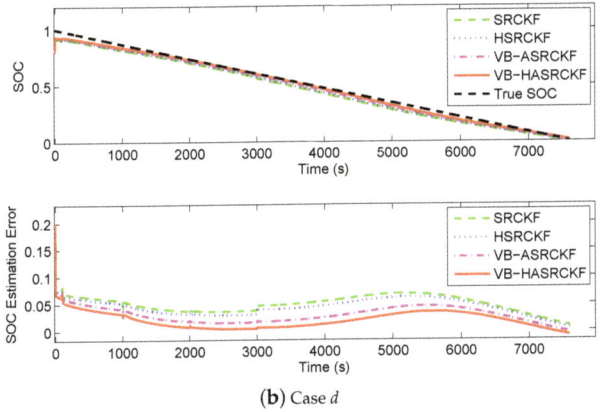

(**b**) Case d

Figure 8. SOC estimation results of four filters in a constant-current discharge test under cases c and d.

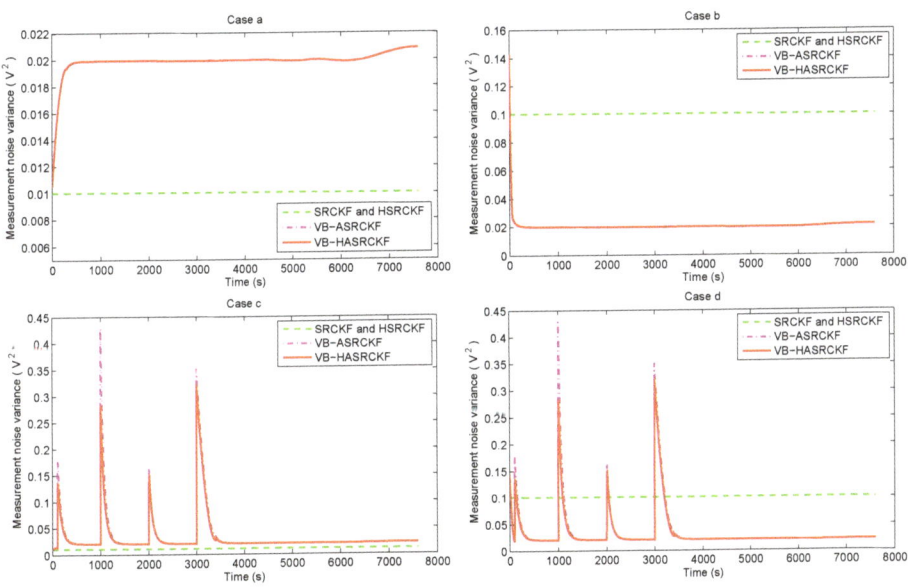

Figure 9. The measurement noise variances in a constant-current discharge test under four cases.

Table 1. The maximum and mean absolute estimation errors of four filters in a constant-current discharge test under four cases. MaE = maximum absolute error, MeE = mean absolute error.

	Case a		Case b		Case c		Case d	
	MaE	MeE	MaE	MeE	MaE	MeE	MaE	MeE
SRCKF	1.52%	0.69%	5.51%	2.24%	6.92%	3.50%	8.15%	4.84%
HSRCKF	1.84%	0.76%	5.51%	2.24%	6.68%	2.47%	7.76%	4.25%
VB-ASRCKF	2.00%	0.67%	2.02%	0.67%	7.00%	2.85%	7.30%	2.85%
VB-HASRCKF	1.97%	0.70%	2.01%	0.70%	6.62%	1.91%	6.73%	1.91%

As given above, the measurement noise variances of SRCKF and HSRCKF were mistuned in case b, and the initial estimates of the measurement variances of VB-ASRCKF and VB-HASRCKF are also mistuned. From the estimation results shown in Figure 7b, we can see that the SOC estimation accuracy of SRCKF and HSRCKF are dramatically declined compared with case a, but VB-ASRCKF and VB-HASRCKF are almost not affected by the mistuning. The estimated measurement noise variances, as shown in Figure 9, still maintain at about 0.02. This illustrates that the VB-based methods have good adaptivity to modeling error by simultaneously estimating the measurement noise variance with the SOC, while SRCKF and HSRCKF have no ability to handle this well.

In case c, some outliers in the voltage and current measurements appeared. The SOC estimation results are presented in Figure 8a. It is clear that VB-HASRCKF and HSRCKF outperform the other two filters in SOC estimation accuracy. It demonstrates that VB-HASRCKF and HSRCKF exhibit robustness to the outliers. In addition, VB-ASRCKF has a more accurate SOC estimation than SRCKF, probably due to the VB method being able to improve the robustness a little bit through the adaptation of the measurement noise variance. This is revealed by the estimated measurement noise variances of VB-ASRCKF and VB-HASRCKF, as given in Figure 9, which are dramatically increased when outliers emerge.

More seriously, in case d, there is both mistuned measurement noise variance and outliers in the measurements. The SOC estimation results are shown in Figure 8b. Clearly, VB-HASRCKF gives the best performance and SRCKF exhibits the worst performance. It can be explained by VB-HASRCKF inheriting the virtue of robustness of Huber's M-estimation and the adaptivity of the VB method. Therefore, the proposed VB-HASRCKF is the most robust and adaptive among these filters.

In addition, the computation times of the four filters under constant-current discharge test are shown in Table 2. It is clear that VB-HASRCKF has the maximum time of computation, and SRCKF has the minimum time of computation. But the difference is not obvious. That is, the proposed VB-HASRCKF shows performance improvement at the cost of a little higher computation complexity.

Table 2. Computation time of the four filters under different tests.

	Constant Current Discharge Test	UDDS Test
SRCKF	12.1816 s	30.1736 s
HSRCKF	12.4577 s	31.1709 s
VB-ASRCKF	13.1058 s	32.7837 s
VB-HASRCKF	14.1003 s	35.4377 s

5.3. SOC Estimation Experimental Results under UDDS Test

To evaluate the SOC estimation performance under dynamic loading profiles, a typical driving cycle, urban dynamometer driving schedule (UDDS), was performed on the battery cell. According to the actual tolerable currents of the lithium-ion battery cell, the loading currents were scaled down, as shown in Figure 10. The maximum current is 3.35A, and the minimum current is 0.01A. The initial SOC value was set to 0.8. The process noise covariance was set as $Q_k = 1 \times 10^{-8} I_2$. The measurement noise variances used for SRCKF and HSRCKF were both set as $R_k = 0.01$ in cases a and c but mistuned to $R_k = 0.1$ in cases b and d. The initial parameter values of VB-based filters were $v_0 = 10$, $V_0 = 0.1$ in cases a and c, while $v_0 = 10$, $V_0 = 1$ were assumed in cases b and d. As can be seen in Figure 11, the outliers in voltage measurements were added at time k = 500 s, 1000 s, 1500 s, 6200 s, 11,000 s, and 15,600 s, with a duration of 5–15 s in cases c and d.

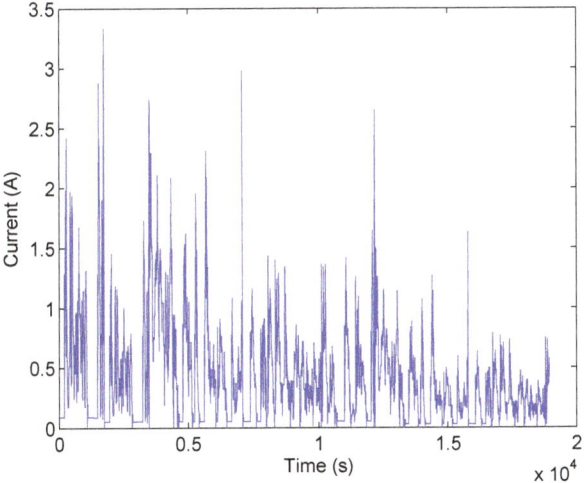

Figure 10. Current profiles in the UDDS test.

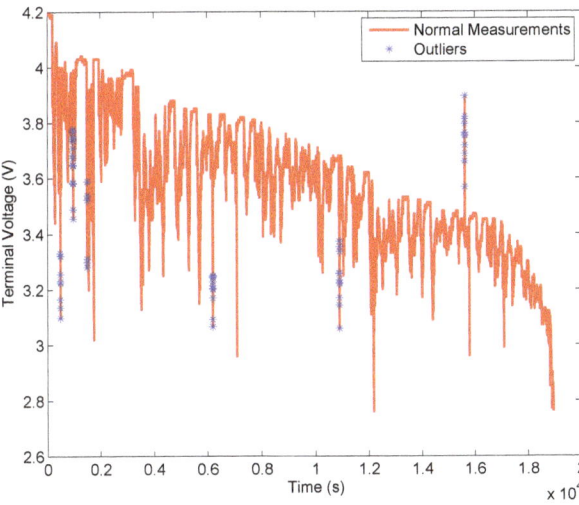

Figure 11. The outlying voltage and current measurements in the UDDS test under cases *c* and *d*.

Figures 12 and 13 present the SOC estimation results of the four filters under four cases. The maximum and mean absolute SOC estimation errors of the four filters are shown in Table 3. Figure 14 depicts the assumed and estimated measurement noise variances of the four filters. It can be seen that the estimated measurement noise variances of VB-ASRCKF and VB-HASRCKF are closer to the assumed values of SRCKF and HSRCKF in case *a*. It shows that the assumed measurement noise variance is consistent with the actual value. Thus, no significant difference occurs in SOC estimation accuracy between the four filters in case *a*. But in case *b*, VB-ASRCKF and VB-HASRCKF have more accurate SOC estimations than SRCKF and HSRCKF due to the adaptivity of the VB methods.

In case *c*, HSRCKF and VB-HASRCKF exhibit higher accuracy than the other two filters. It means that Huber's M-estimation based methods are more robust to outliers. Moreover, as given in Figure 14, the estimated measurement noise variances of VB-ASRCKF and VB-HASRCKF increase notably when an outlier appears. It illustrates that the VB-based filter can adapt the measurement noise variance online according to the actual situation. This is why VB-ASRCKF has more accurate SOC estimation than SRCKF in the presence of outliers. In case d, VB-HASRCKF has the most accurate SOC estimation. The mistuned measurement noise variance and outliers in measurements have a negligible effect on VB-HASRCKF, while the effect is obvious for the other three filters.

The computation times of the four filters under the UDDS test are shown in Table 2. It can be seen that the computation time of VB-HASRCKF is 35.4377 s, while it is 30.1736 s for SRCKF. It shows that VB-HASRCKF has a little higher computation complexity.

In a word, by these comparison results, we can conclude that the proposed VB-HASRCKF algorithm has more accurate and more stable SOC estimation than the other three filters at the price of a little more computation time.

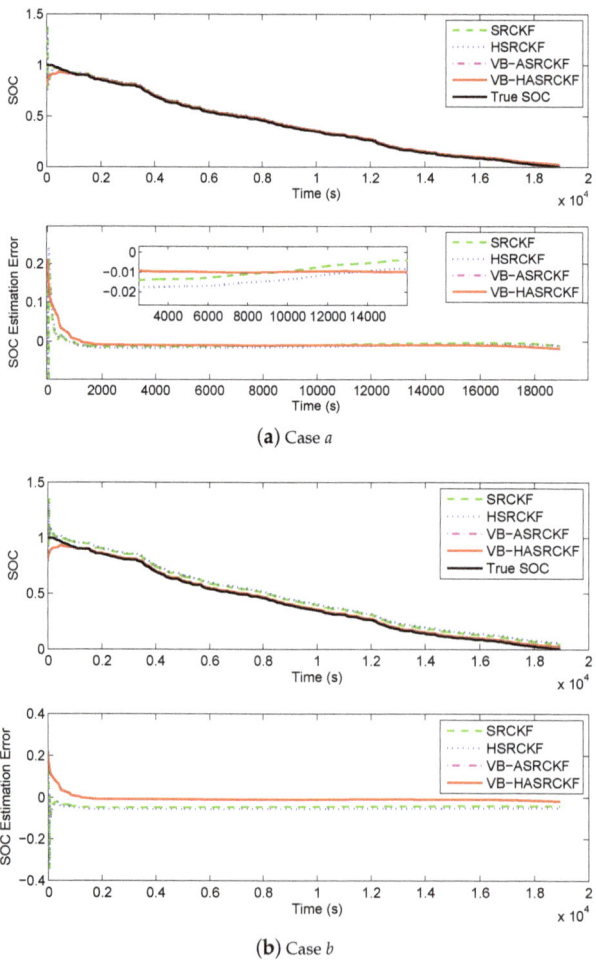

Figure 12. SOC estimation results of four filters in the UDDS test under cases a and b.

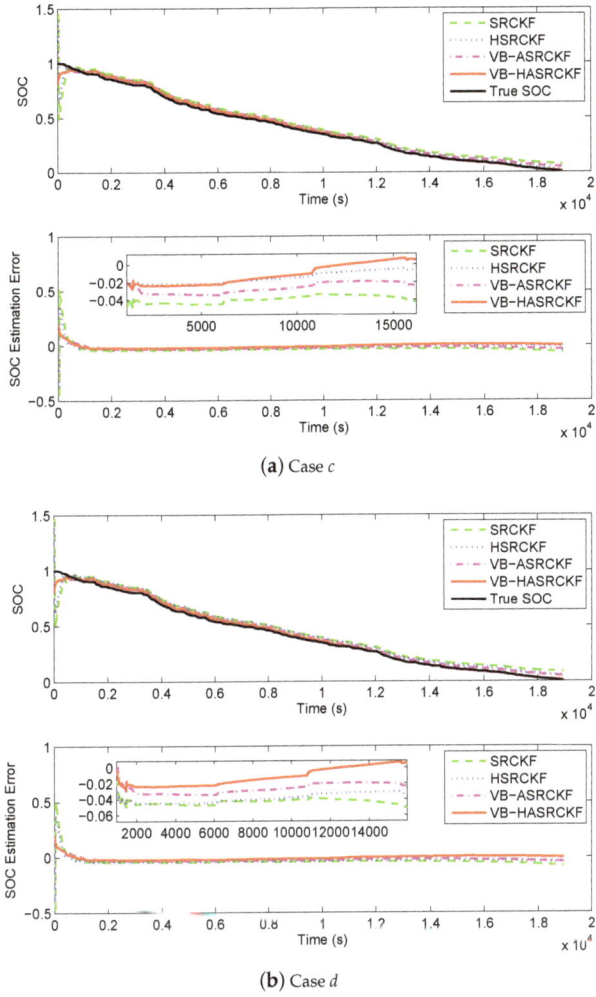

Figure 13. SOC estimation results of four filters in the UDDS test under cases *c* and *d*.

Table 3. The maximum and mean absolute estimation errors of four filters (after 10 min) in the UDDS test under four cases.

	Case *a*		Case *b*		Case *c*		Case *d*	
	MaE	MeE	MaE	MeE	MaE	MeE	MaE	MeE
SRCKF	1.46%	0.91%	4.87%	4.49%	6.72%	4.14%	7.88%	4.63%
HSRCKF	1.80%	1.28%	5.68%	5.38%	2.49%	1.44%	4.77%	3.85%
VB-ASRCKF	2.85%	1.03%	2.88%	1.03%	4.15%	2.71%	4.22%	2.71%
VB-HASRCKF	2.85%	1.03%	2.84%	1.03%	2.80%	1.18%	2.96%	1.18%

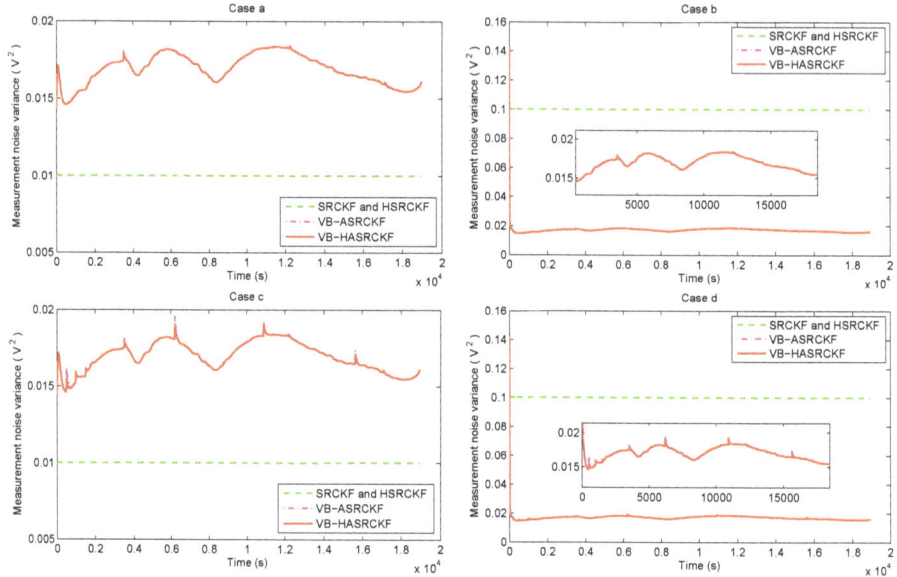

Figure 14. The measurement variances in the UDDS test under four cases.

6. Conclusions

Aiming to improve the accuracy and robustness of the SOC estimation of the lithium-ion battery under the condition of unknown or time-varying measurement noise covariance and outliers in the measurements, an adaptive and robust square root cubature Kalman filter based on variational Bayesian approximation and Huber's M-estimation (VB-HASRCKF) is proposed in this paper. The measurement noise covariance is simultaneously estimated with the SOC to account for battery model uncertainties and measurement noise covariance uncertainties, reducing the impact of model mismatch. Meanwhile, the outlying current and voltage measurements caused by adverse operating conditions are accounted for by Huber's M-estimation. Through experiments under a constant-current discharge test and a UDDS test, the effectiveness and superiority of the proposed algorithm were verified by comparison with SRCKF, HSRCKF, and VB-ASRCKF in terms of SOC estimation accuracy and robustness. Especially when there is both mistuned measurement noise covariance and outliers in the measurements, the proposed VB-HASRCKF exhibits significant performance superiority, although with a little higher computation complexity.

In addition, some comments should be made about the performance comparison with the data-driven method. As mentioned in [12], the data-driven method may be divergent with bad parameter selection when the training data cannot completely cover the present operating conditions. That is to say, if the training data does not contain the outliers in current and voltage measurements and the parameters are not appropriately chosen, the data-driven method may not exhibit as good a performance in SOC estimation accuracy as our proposed method. If the training sample set considers the case of outliers, it may have better performance than our proposed method. Even in this case, it does not mean that our method is meaningless. There is no need for a lot of training data and a large amount of computation for our method, which in itself is an advantage over the data-driven method. Of course, we will undertake a deeper comprative study on the performance of the model-based method and the data-driven method based on enough experimental data in our future work. In addition, a combination of the data-driven method and the Kalman filter-based method needs to be further studied so as to develop their strengths and compensate their weaknesses. Uncertainties in battery capacity should also be particularly considered in future work to account for battery aging.

Author Contributions: J.H. conceived this paper, designed the experiments, and analyzed the data; H.H. did the investigation; T.G. and Y.Z. performed the experiments; Y.Y. revised the paper and provided some valuable suggestions.

Funding: This research was funded by the Fundamental Research Funds for the Central Universities (No. 3102019ZX020), Shaanxi Provincial Key Research and Development Programs (No. 2017ZDXM-GY-06, No. 2017GY-057, 2019GY-003), and Xi'an Science and Technology Planning Project–Scientific and Technological Innovation Guidance Project (No. 201805042YD20CG26 (8)).

Conflicts of Interest: The authors declare no conflict of interest.

Abbreviations

The following abbreviations are used in this manuscript:

ACKF	Adaptive Cubature Kalman Filter
ALS	Autocovariance Least Squares
ASRCKF	Adaptive Square Root Cubature Kalman Filter
ANN	Artificial Neural Network
BESS	Battery Energy Storage System
BMS	Battery Management System
CC	Coulomb Counting
CE	Coulomb Efficiency
CKF	Cubature Kalman Filter
ECM	Equivalent Circuit Model
EKF	Extended Kalman Filter
EV	Electric Vehicle
FL	Fuzzy Logic
IW	Inverse Wishart
KF	Kalman Filter
KL	Kullback–Leibler
MM	Multiple Model
OCV	Open-Circuit Voltage
PF	Particle Filter
RC	Resistor–Capacitor
SOC	State Of Charge
SVM	Support Vector Machine
SRCKF	Square Root Cubature Kalman Filter
UDDS	Urban Dynamometer Driving Schedule
UKF	Unscented Kalman Filter
VB	Variational Bayesian

References

1. Yu, Q.; Xiong, R.; Lin, C.; Shen, W.; Deng, J. Lithium-ion battery parameters and state-of-charge joint estimation based on H-infinity and unscented Kalman filters. *IEEE Trans. Veh. Technol.* **2017**, *66*, 8693–8701. [CrossRef]
2. Huang, C.; Wang, Z.; Zhao, Z.; Wang, L.; Lai, C.S.; Wang, D. Robustness Evaluation of Extended and Unscented Kalman Filter for Battery State of Charge Estimation. *IEEE Access* **2018**, *6*, 27617–27628. [CrossRef]
3. He, W.; Williard, N.; Chen, C.; Pecht, M. State of charge estimation for Li-ion batteries using neural network modeling and unscented Kalman filter-based error cancellation. *Int. J. Elect. Power Energy Syst.* **2014**, *62*, 783–791. [CrossRef]
4. Zhao, W.; Kong, X.; Wang, C. Combined estimation of the state of charge of a lithium battery based on a back-propagation- adaptive Kalman filter algorithm. *Proc. Inst. Mech. Eng. Part D* **2018**, *232*, 357–366. [CrossRef]

5. Charkhgard, M.; Farrokhi, M. State-of-charge estimation for lithium-ion batteries using neural networks and EKF. *IEEE Trans. Ind. Electron.* **2010**, *57*, 4178–4187. [CrossRef]
6. Singh, P.; Vinjamuri, R.; Wang, X.; Reisner, D. Design and implementation of a fuzzy logic-based state-of-charge meter for Li-ion batteries used in portable defibrillators. *J. Power Sources* **2006**, *162*, 829–836. [CrossRef]
7. Cai, C.H.; Du, D.; Liu, Z.Y. Battery state-of-charge (SOC) estimation using adaptive neuro-fuzzy inference system (ANFIS). In Proceedings of the 12th IEEE International Conference on Fuzzy Systems, St Louis, MO, USA, 25–28 May 2003; Volume 2, pp. 1068–1073.
8. Awadallah, M.A.; Venkatesh, B. Accuracy improvement of SOC estimation in lithium-ion batteries. *J. Energy Storage* **2016**, *6*, 95–104. [CrossRef]
9. Hu, J.N.; Hu, J.J.; Lin, H.B.; Li, X.P.; Jiang, C.L.; Qiu, X.H.; Li, W.S. State-of-charge estimation for batterymanagement system using optimized support vectormachine for regression. *J. Power Sources* **2014**, *269*, 682–693. [CrossRef]
10. Sheng, H.; Xiao, J. Electric vehicle state of charge estimation: Nonlinear correlation and fuzzy support vector machine. *J. Power Sources* **2015**, *281*, 131–137. [CrossRef]
11. Wu, X.; Mi, L.; Tan, W.; Qin, J.L.; Zhao, M.N. State of charge (SOC) estimation of Ni-MH battery based on least square support vector machines. *Adv. Mater. Res.* **2011**, *211*, 1204–1209. [CrossRef]
12. Xiong, R.; Cao, J.; Yu, Q.; He, H.; Sun, F. Critical review on the battery state of charge estimation methods for electric vehicles. *IEEE Access* **2018**, *6*, 1832–1843. [CrossRef]
13. Xu, J.; Mi, C.C.; Cao, B.; Deng, J.; Chen, Z.; Li, S. The state of charge estimation of lithium-ion batteries based on a proportional-integral observer. *IEEE Trans. Veh. Technol.* **2014**, *63*, 1614–1621.
14. Huangfu, Y.; Xu, J.; Zhao, D.; Liu, Y.; Gao, F. A novel battery state of charge estimation method based on a super-twisting sliding mode observer. *Energies* **2018**, *11*, 1211. [CrossRef]
15. Plett, G.L. Extended Kalman filtering for battery management systems of LiPB-based HEV battery packs: Part 3. State and parameter estimation. *J. Power Sources* **2004**, *134*, 277–292. [CrossRef]
16. He, W.; Williard, N.; Chen, C.; Pecht, M. State of charge estimation for electric vehicle batteries using unscented kalman filtering. *Microelectron. Reliab.* **2013**, *53*, 840–847. [CrossRef]
17. Li, Y.; Wang, C.; Gong, J. A multi-model probability SOC fusion estimation approach using an improved adaptive unscented Kalman filter technique. *Energy* **2017**, *141*, 1402–1415. [CrossRef]
18. Peng, S.; Chen, C.; Shi, H.; Yao, Z. State of charge estimation of battery energy storage systems based on adaptive unscented Kalman filter with a noise statistics estimator. *IEEE Access* **2017**, *5*, 13202–1312. [CrossRef]
19. Wang, X.; Song, Z.; Yang, K.; Yin, X.; Geng, Y.; Wang, J. State of Charge Estimation for Lithium-Bismuth Liquid Metal Batteries. *Energies* **2019**, *12*, 183. [CrossRef]
20. Xia, B.; Wang, H.; Tian, Y.; Wang, M.; Sun, W.; Xu, Z. State of charge estimation of lithium-ion batteries using an adaptive cubature Kalman filter. *Energies* **2015**, *8*, 5916–5936. [CrossRef]
21. Zeng, Z.; Tian, J.; Li, D.; Tian, Y. An online state of charge estimation algorithm for lithium-ion batteries using an improved adaptive cubature Kalman filter. *Energies* **2018**, *11*, 59. [CrossRef]
22. Cui, X.; Jing, Z.; Luo, M.; Guo, Y.; Qiao, H. A New Method for State of Charge Estimation of Lithium-Ion Batteries Using Square Root Cubature Kalman Filter. *Energies* **2018**, *11*, 209. [CrossRef]
23. Chen, L.; Xu, L.; Wang, R. State of Charge Estimation for Lithium-Ion Battery by Using Dual Square Root Cubature Kalman Filter. *Math. Probl. Eng.* **2017**, *2017*, 5489356. [CrossRef]
24. Liu, S.; Cui, N.; Zhang, C. An adaptive square root unscented Kalman filter approach for state of charge estimation of lithium-ion batteries. *Energies* **2017**, *10*, 1345.
25. El Din, M.S.; Abdel-Hafez, M.F.; Hussein, A.A. Enhancement in Li-ion battery cell state-of-charge estimation under uncertain model statistics. *IEEE Trans. Veh. Technol.* **2016**, *65*, 4608–4618. [CrossRef]
26. Charkhgard, M.; Zarif, M.H. Design of adaptive H∞ filter for implementing on state of- charge estimation based on battery state-of-charge-varying modelling. *IET Power Electron.* **2015**, *8*, 1825–1833. [CrossRef]
27. Zhao, L.; Liu, Z.; Ji, G. Lithium-ion battery state of charge estimation with model parameters adaptation using H∞ extended Kalman filter. *Control Eng. Pract.* **2018**, *81*, 114–128. [CrossRef]
28. Wei, Z.; Leng, F.; He, Z.; Zhang, W.; Li, K. Online state of charge and state of health estimation for a Lithium-Ion battery based on a data-model fusion method. *Energies* **2018**, *11*, 1810. [CrossRef]
29. Sarkka, S.; Nummenmaa, A. Recursive noise adaptive Kalman filtering by variational Bayesian approximations. *IEEE Trans. Automat. Contr.* **2009**, *54*, 596–600. [CrossRef]

30. Li, K.; Chang, L.; Hu, B. A variational Bayesian-based unscented Kalman filter with both adaptivity and robustness. *IEEE Sens. J.* **2016**, *16*, 6966–6976. [CrossRef]
31. Sun, J.; Zhou, J.; Li, X. R. State estimation for systems with unknown inputs based on variational Bayes method. In Proceedings of the 15th International Conference on Information Fusion, Singapore, 9–12 July 2012; pp. 983–990.
32. Hou, J.; Yang, Y.; Gao, T. Variational Bayesian based adaptive shifted Rayleigh filter for bearings-only tracking in clutters. *Sensors* **2019**, *19*, 1512. [CrossRef]
33. Hou, J.; Yang, Y.; He, H.; Gao, T. Adaptive Dual Extended Kalman Filter Based on Variational Bayesian Approximation for Joint Estimation of Lithium-Ion Battery State of Charge and Model Parameters. *Appl. Sci.* **2019**, *9*, 1726. [CrossRef]
34. Chang, L.; Hu, B.; Chang, G.; Li, A. Huber-based novel robust unscented Kalman filter. *IET Sci. Meas. Technol.* **2012**, *6*, 502–509. [CrossRef]
35. He, W.; Pecht, M.; Flynn, D.; Dinmohammadi, F. A Physics-Based Electrochemical Model for Lithium-Ion Battery State-of-Charge Estimation Solved by an Optimised Projection-Based Method and Moving-Window Filtering. *Energies* **2018**, *11*, 2120. [CrossRef]
36. Yang, F.; Wang, D.; Zhao, Y.; Tsui, K.L.; Bae, S. J. A study of the relationship between coulombic efficiency and capacity degradation of commercial lithium-ion batteries. *Energy* **2018**, *145*, 486–495. [CrossRef]
37. Zheng, Y.; Ouyang, M.; Lu, L.; Li, J.; Zhang, Z.; Li, X. Study on the correlation between state of charge and coulombic efficiency for commercial lithium ion batteries. *J. Power Sources* **2015**, *289*, 81–90. [CrossRef]
38. Wang, J.; Cao, B.; Chen, Q.; Wang, F. Combined state of charge estimator for electric vehicle battery pack. *Control Eng. Pract.* **2007**, *15*, 1569–1576. [CrossRef]
39. Smith, A.J.; Burns, J.C.; Dahn, J.R. A high precision study of the Coulombic efficiency of Li-ion batteries. *Electrochem. Solid-State Lett.* **2010**, *13*, A177–A179. [CrossRef]
40. Wang, Q.; Kang, J.; Tan, Z.; Luo, M. An online method to simultaneously identify the parameters and estimate states for lithium ion batteries. *Electrochim. Acta* **2018**, *289*, 376–388. [CrossRef]

© 2019 by the authors. Licensee MDPI, Basel, Switzerland. This article is an open access article distributed under the terms and conditions of the Creative Commons Attribution (CC BY) license (http://creativecommons.org/licenses/by/4.0/).

Article

A Coupled-Inductor DC-DC Converter with Input Current Ripple Minimization for Fuel Cell Vehicles

Fuwu Yan [1], Jingyuan Li [1], Changqing Du [1], Chendong Zhao [2], Wei Zhang [3] and Yun Zhang [3],*

1. Hubei Key Laboratory of Advanced Technology for Automotive Components, Wuhan University of Technology, Wuhan 430070, China; yanfuwu@vip.sina.com (F.Y.); lijingyuan@catarc.ac.cn (J.L.); cq_du@whut.edu.cn (C.D.)
2. State Grid Shanghai Jiading Electric Power Supply Company, Shanghai 201800, China; zhaochendong@tju.edu.cn
3. School of Electrical and Information Engineering, Tianjin University, Tianjin 300070, China; duanyidelei@163.com
* Correspondence: zhangy@tju.edu.cn

Received: 16 March 2019; Accepted: 23 April 2019; Published: 5 May 2019

Abstract: A coupled-inductor DC-DC converter with a high voltage gain is proposed in this paper to match the voltage of a fuel cell stack to a DC link bus. The proposed converter can minimize current ripples and can also achieve a high voltage gain by adjusting the duty cycle d and the turns ratio n of a coupled inductor. A passive lossless clamping circuit that is composed of one capacitor and one diode is employed, and this suppresses voltage spikes across the power device resulting from system leakage inductance. The operating principles and the characteristics of the proposed converter are analyzed and discussed. A 400-W experimental prototype was developed, and it had a wide voltage gain range (4–13.33) and a maximum efficiency of 95.12%.

Keywords: coupled inductor; DC-DC converter; high voltage gain; ripple minimization current; fuel cell vehicles

---

1. Introduction

With extensive use of fossil fuels in transport, power, and heating, addressing the imminent global energy crisis has become an increasing concern [1–3]. In addition, considerations such as the greenhouse effect and increasing air pollution (especially in cities) caused particularly by emissions from fossil fuel vehicles have become significant influences on quality of life and health [4]. Renewable energy vehicles, including fuel cell vehicles, pure electric vehicles, and hybrid energy source vehicles [5,6], can greatly reduce the impact of transport on the environment due to their pollution-free characteristics [7,8]. Fuel cell vehicles have attracted wide attention because of their higher energy conversion rate compared to locomotives [9] and their longer cycle range compared to battery electric vehicles [10]. However, a fuel cell source cannot be connected to the propulsion system's DC bus directly, as it can have a low and variable output voltage and a large output current [11]. As a result, a DC-DC boost converter is needed to lift the voltage of the fuel cell source and to match the DC bus voltage [12]. Above all, a DC-DC converter with a minimal input current ripple and a high voltage gain is extremely important for the future development of fuel cell vehicles.

The traditional DC-DC boost converter features a simple structure and low component counts. However, low voltage gain and high voltage stresses across all semiconductor devices limit its application. Although a three-level DC-DC boost converter lowers the voltage stress to half of the output voltage, the duty cycle is extreme, while the voltage gain is high. A three-level DC-DC converter with a high voltage gain and without extreme duty cycles was proposed in Reference [13], but it had a greater cost due to the number of semiconductors (eight MOSFETs and four diodes), and the control strategy was also more complicated. In References [14,15], the proposed converters employed switched-capacitor cells to achieve a high voltage gain, while the number of switched-capacitor cells, the volume of the converter, and the cost increased. Coupled inductors have been employed to improve the voltage gain of converters (proposed in [16,17]). However, the higher the voltage gain was, the greater the turns ratio required from the coupled inductors was, which increased the leakage inductance of the coupled inductors and made manufacturing more difficult. A novel high step-up DC-DC converter with coupled inductors and a switched capacitor was proposed in Reference [18]. However, the leakage inductance of the coupled inductors resulted in a high voltage spike across the power switches. An active clamp [19] or a passive snubber circuit [20] are often adopted to solve the aforementioned voltage spike issues. However, the complexity of the control strategy and the circuit also increase. At the same time, these converters have different levels of input current ripples, and a large current ripple has a serious impact on the lifetime of fuel cells [21]. In References [22–24], interleaved converters were proposed to cut down the input current ripple. The improved modulation strategy proposed in Reference [13] could also obtain a lower input current ripple. However, as the voltage gain increased, the input current became large, and therefore the increased current ripple was still a disadvantage for fuel cell sources. An active compensation method was proposed in Reference [25] based on the designed push–pull DC-DC converter by adding an active-clamp circuit to further reduce current ripples in the uninterruptible power supply (UPS) system. However, the method focused on the low-frequency current ripple and employed an extra circuit. A maximum power point tracking (MPPT) control with a perturbed duty ratio D_p was employed in Reference [26] to simply reduce the input ripple current of a proton exchange membrane (PEM) fuel cell. However, it was only suitable for AC output applications. A pre-regulator was introduced in Reference [27] to interface with the main regulator for reducing the current ripple. However, the pre-regulator employed an interleaved structure that was not indispensable. The introduced pre-regulator reduced the power density and increased the cost of the proposed converter. To reduce the low-frequency input current ripple without the auxiliary circuit, a control method based on a front-end DC-DC converter was proposed in Reference [28]. This method intended to modify the DC bus voltage reference and control the DC bus voltage to fluctuate properly at $2f_o$, making the DC bus capacitor support nearly all of the fluctuating power. However, it was not suitable for high-frequency current ripples caused by a charging/discharging current flowing through the input inductor.

Based on the step-up DC-DC converter with an input current ripple minimization proposed in Reference [29], a novel coupled-inductor converter with a high voltage gain is proposed in this paper. The converter benefits from input current ripple minimization, further improves voltage gain, and reduces voltage stresses. A passive lossless clamping circuit is introduced, consisting of one capacitor and one diode. As a result, voltage spikes across the power switch caused by leakage inductance can be suppressed. In addition, the output and the input share a common ground, which can eliminate additional electromagnetic interference (EMI) issues. Section 2 presents the topology of the proposed converter. The operating principles and characteristics of the converter are analyzed in Section 3 in detail. In Section 4, the voltage and current stresses of the semiconductors are analyzed, and a comparison to the counterpart in Reference [29] is made. In Section 5, an experimental prototype is built, and experimental results validate the theoretical analysis.

2. Topology

The topology of the proposed converter and its equivalent circuit are shown in Figure 1a,b, respectively. As the equivalent circuit shows, the coupled inductor is composed of a magnetizing inductance L_M, a leakage inductance L_r, and an ideal transformer whose turns ratio is $n_p:n_s = 1:n$. The inductor L_a and the capacitor C_1 form an input current ripple minimization unit, and the diode D_1 and the capacitor C_4 form a passive lossless clamping circuit. Consequently, the voltage stress across the power switch Q can be clamped to the voltage across C_4 when Q is turned off. The voltage doubling unit is composed of the capacitor C_2 and the coupled inductor L_s, while the capacitor C_3 and the capacitor C_4 are the energy storage capacitors in the high voltage side.

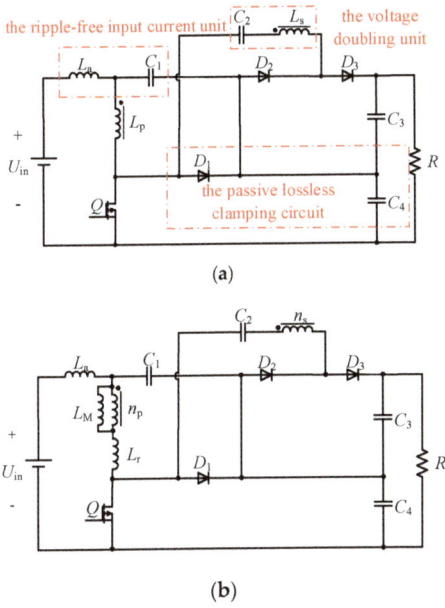

Figure 1. The proposed coupled-inductor DC-DC converter. (a) The topology; (b) the equivalent circuit.

3. Analysis of Operating Principles and Characteristics

3.1. Operating Principles

The proposed converter has 10 operating modes (in which Mode G and Mode H cannot coexist simultaneously in a single switching period), and the corresponding current flow paths are depicted in Figure 2.

Mode A: The power switch Q is turned on. The currents that flow through both sides of the ideal transformer decrease rapidly. Meanwhile, diode D_3 is still conducting, and the current flowing through capacitor C_1 starts to decrease. This mode ends when the current flowing in C_3 becomes zero.

Mode B: The current flowing into diode D_3 is no longer able to provide energy for the load and continues to decrease. Accordingly, capacitor C_3 begins to discharge, and the current flowing through it increases. The trends of the currents flowing in other branches of the circuit remain the same. This mode ends when the current flowing through diode D_3 decreases to zero.

Mode C: The currents flowing into both sides of the ideal transformer are reversed and gradually increase. Diode D_2 is turned on, and capacitor C_3 is still discharging. The coupled inductor transfers energy to capacitor C_2, and it starts to charge. The capacitor C_1 continues to discharge, but its current decreases gradually. Meanwhile, capacitor C_4 continues to charge, and its current gradually decreases to zero, at which time this mode ends.

Mode D: The current flowing through capacitor C_4 begins to reverse and gradually increase, and C_4 begins to discharge. The current flowing through capacitor C_1 continuously decreases, while the current flowing in diode D_2 rises gradually. This mode ends when the current flowing through capacitor C_1 is equal to the current flowing in diode D_2.

Mode E: The current flowing in diode D_2 continues to increase, and it is larger than the current flowing through capacitor C_1. As a result, the current flowing through capacitor C_4 continues to rise. When the current flowing into capacitor C_1 decreases to zero, this mode ends.

Mode F: Capacitor C_1 starts to charge, and the current flowing in it increases gradually. Then the current decreases as the capacitor voltage rises.

Mode G: The charging process of capacitor C_2 is completed, and the current flowing in the secondary side of the ideal transformer n_s reduces to zero. The current flowing in the primary side n_p of the ideal transformer increases faster because there is no energy being transferred to the secondary side. This mode does not exist when the duty cycle is small (based on the magnetizing inductor).

Mode H: This mode and Mode G are mutually exclusive, and the converter goes directly into Mode I when Mode G exists. Q is turned off at the beginning, and the magnetizing inductor L_M discharges. As a result, the current flowing into the leakage inductance L_r starts to decrease. Because of the conduction of diode D_1, the voltage stress across Q is clamped at the voltage across the capacitor C_4. The capacitor C_4 continues charging. Meanwhile, capacitor C_2 is still in the charging state because the currents flowing into both sides of the ideal transformer have not reversed yet. The diode D_2 remains turned on, and the current flowing through capacitor C_1 begins to decrease. This mode finishes when the current flowing in capacitor C_2 reduces to zero.

Mode I: In this mode, the currents flowing into both sides of the ideal transformer reverse and start to increase, and the current flowing into C_3 begins to decrease because of the conduction of D_3. When the current flowing into C_3 decreases to zero, this mode ends.

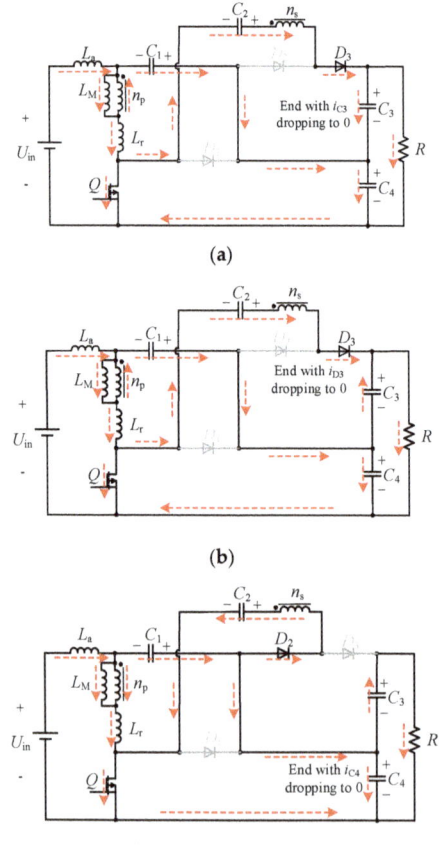

Figure 2. *Cont.*

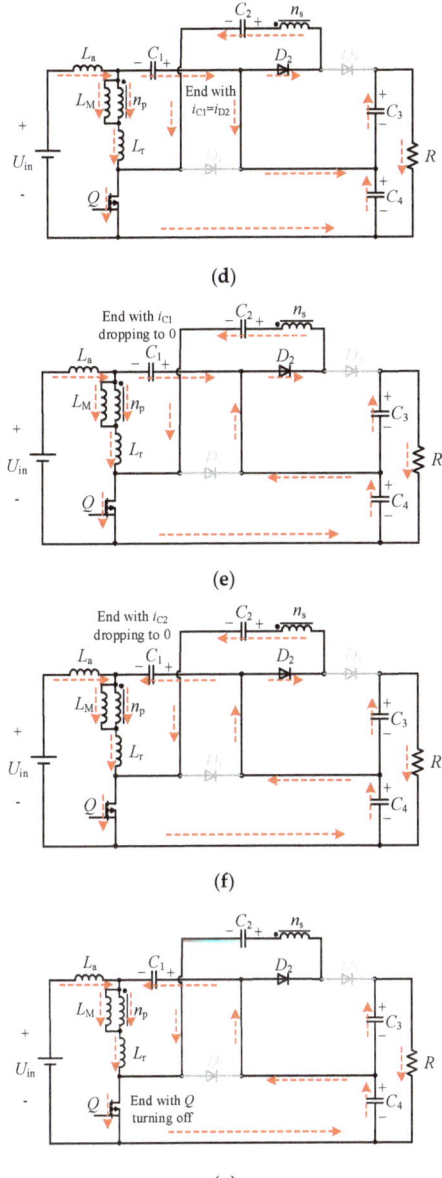

Figure 2. Cont.

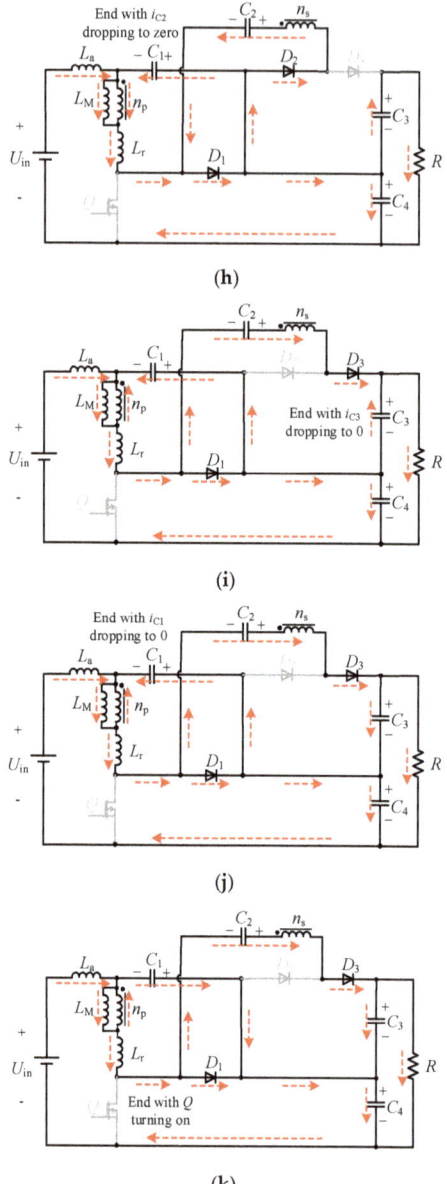

Figure 2. Current flow paths of the proposed converter in one switching period. (**a**) Mode A; (**b**) Mode B; (**c**) Mode C; (**d**) Mode D; (**e**) Mode E; (**f**) Mode F; (**g**) Mode G; (**h**) Mode H; (**i**) Mode I; (**j**) Mode J; (**k**) Mode K.

Mode J: The current flowing into D_3 continues to increase, and C_3 changes into a charging state. The current trends of other branches remain unchanged. This mode terminates when the current flowing into C_1 decreases to zero.

Mode K: C_1 begins to discharge, and energy is transferred to C_4 from C_1. As the current flowing through L_a does not change, the current flowing into the primary side n_p falls. As a result, the current flowing into the secondary side n_s begins to decrease. This mode terminates when the power switch Q turns on.

Mode F, Mode G, and Mode K have the longest durations of these 10 modes (Mode G and Mode H cannot coexist simultaneously), while the other 7 modes are transient states that last for a very short time. As a result, the simplified current waveforms of the circuit elements in a switching period are given in Figure 3.

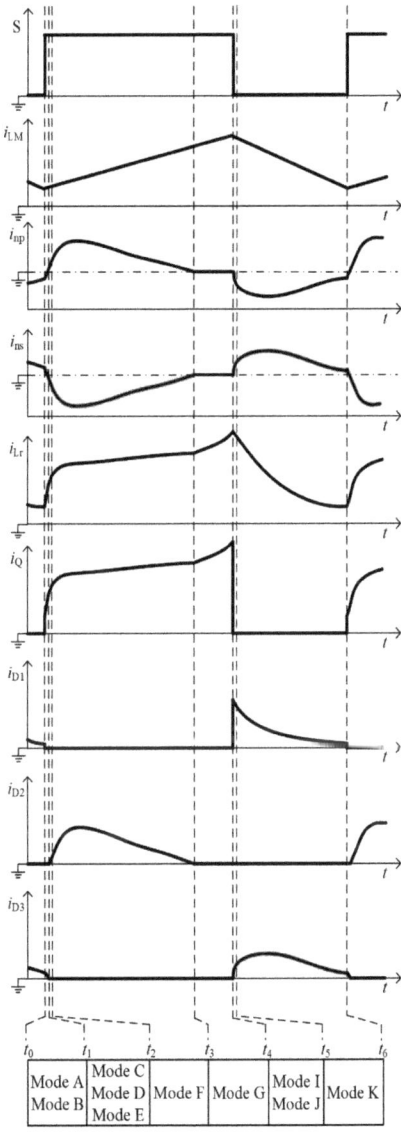

Figure 3. The simplified current waveforms of the converter. Mode A: End with i_{C3} dropping to 0; Mode B: End with i_{D3} dropping to 0; Mode C: End with i_{C4} dropping to 0; Mode D: End with i_{C1} equaling i_{D2}; Mode E: End with i_{C1} dropping to 0; Mode I: End with i_{C3} dropping to 0; Mode J: End with i_{C1} dropping to 0.

According to the states of Mode A–Mode G shown in Figure 2, the relationship between inductors and a current ripple can be obtained as

$$L_M + L_a + L_r = \frac{U_{in}d}{2f\alpha I_L} \tag{1}$$

where α is the current ripple coefficient. The output filtering capacitors are chosen to be large enough that the voltage ripples can be limited well. Therefore, the relationship between C_4 and its voltage ripple can be obtained as

$$C_4 \geq \frac{\Delta i_L T_s}{8\Delta u} = \frac{\alpha I_L T_s}{8\Delta u} \tag{2}$$

3.2. Analysis of Ripple Minimization Characteristics

Assume that the capacitors and the inductors employed are large enough, and the forward voltage and the on-state resistances of all semiconductors are negligible. The average voltages across C_1, C_2, C_3, and C_4 are U_{C1}, U_{C2}, U_{C3}, and U_{C4} respectively; U_{in} is the steady input voltage; u_{La} is the instantaneous inductor voltage; the instantaneous current flowing into L_a is i_{La} and the average inductor current is I_{La}; and the duty cycle is d.

Over the whole switching period, the input inductor L_a, the capacitor C_1, and the capacitor C_4 form a closed circuit loop, and the following equation can be derived by applying Kirchhoff's Voltage Laws (KVLs):

$$\begin{cases} u_{La} = u_{in} - u_{C4} + u_{C1} \\ u_{La} = L_a \frac{di_{La}(t)}{dt} \end{cases} \tag{3}$$

where u_{in} is the instantaneous input voltage, and u_{c1} and u_{c4} are the instantaneous voltages across C_1 and C_4.

Considering the steady state, the inductor L_a satisfies the voltage second balance principle, and thus the average voltage across the inductor L_a is zero in each switching period. Therefore, Equation (4) can be obtained. The voltage ripples across C_1 and C_4 are very close to zero, and the input voltage is regarded as constant in a switching period, so the instantaneous inductor voltage u_{La} is very small, according to Equation (3). Therefore, the current ripple of the inductor current i_{La} can be regarded as zero, which means that input current ripple minimization can be realized:

$$\begin{cases} U_{C1} = U_{C4} - U_{in} \\ U_{La} = 0 \end{cases} \tag{4}$$

It can be seen that the characteristics of current ripple minimization are affected by L_a, u_{C1}, and u_{C4}. The capacitor voltage fluctuation is smaller if the capacitances of C_1 and C_4 are larger, which is beneficial to the realization of current ripple minimization.

3.3. Voltage Gain Analysis

In a steady state, neglecting the other transient state modes, the main modes (Mode F, Mode G, and Mode K) are analyzed. In view of the influence of leakage inductance on the voltage gain, the coupling coefficient $k = L_M/(L_M + L_r)$ is introduced.

When the power switch Q is turned on, the voltage across the primary side n_p of the ideal transformer is equal to kU_{in}, while the voltage across the secondary side n_s is $U_{C2} - U_{C4}$. Equation (5) can be obtained:

$$nkU_{in} = U_{C4} - U_{C2}. \tag{5}$$

When the power switch Q is turned off, the voltage across the primary side n_p is changed to kU_{C1}, and the voltage across the secondary side n_s becomes $U_{C3} - U_{C2}$. Therefore, Equation (6) can be obtained:

$$nkU_{C1} = U_{C3} - U_{C2}. \tag{6}$$

Applying the voltage second balance principle to the magnetizing inductor L_M yields Equation (7):

$$kU_{in}d = k(U_{C4} - U_{in})(1-d). \tag{7}$$

From Equations (4)–(7), the voltages across C_1, C_2, C_3, and C_4 can be derived as

$$\begin{cases} U_{C1} = \dfrac{d}{1-d}U_{in} \\ U_{C2} = \dfrac{nk - nkd + 1}{1-d}U_{in} \\ U_{C3} = \dfrac{nk + 1}{1-d}U_{in} \\ U_{C4} = \dfrac{1}{1-d}U_{in} \end{cases}. \tag{8}$$

Above all, the voltage gain M of the proposed converter can be expressed as

$$M = \frac{U_{out}}{U_{in}} = \frac{U_{C3} + U_{C4}}{U_{in}} = \frac{nk + 2}{1-d}. \tag{9}$$

3.4. Analysis of Passive Lossless Clamping Circuit

As can be seen in Figure 2h, the diode D_1 and the capacitor C_4 form a passive lossless clamping circuit. C_4 can absorb the energy stored in the leakage inductance through D_1 and limit the voltage stress across the power switch Q to U_{C4}. This passive lossless clamping circuit provides a path for transmitting leakage inductance energy to the load. Moreover, it can raise the efficiency of the converter and suppress the voltage spike across the power switch Q caused by leakage inductance L_r.

4. Voltage and Current Stresses and Performance Comparisons

4.1. Voltage Stresses

In view of the analysis of the operating modes, the power switch Q and the capacitor C_4 are connected in parallel when Q is turned off, and D_2 is connected to the capacitor C_3 in parallel when D_2 is turned off. In other words, the voltages across Q and C_4 are equal, and the voltage stress across D_2 is U_{C3}. The voltage stresses across Q and D_2 can be derived as

$$U_Q = U_{C4} = \frac{1}{1-d}U_{in}, \tag{10}$$

$$U_{D2} = U_{C3} = \frac{nk + 1}{1-d}U_{in}. \tag{11}$$

Similarly, the voltage stress across D_1 is U_{C4} and the voltage stress across D_3 is U_{C3} when Q is turned on. Thus, the voltage stresses across D_1 and D_3 can be derived as

$$U_{D1} = U_{C4} = \frac{1}{1-d}U_{in}, \tag{12}$$

$$U_{D3} = U_{C3} = \frac{nk + 1}{1-d}U_{in}. \tag{13}$$

It can be concluded that the voltage stresses across the power switch Q and the diode D_1 are independent of the coupled inductors and are only related to the duty cycle d and the input voltage U_{in}. The voltage stresses across diode D_2 and diode D_3 are not only related to d, but are also related to the turns ratio n. Hence, it is beneficial to choose a desired turns ratio n for the coupled inductors to obtain a trade-off between the voltage gain and the voltage stresses across D_2 and D_3.

4.2. Current Stresses

To simplify the calculation of current stresses, the influence of leakage inductance is neglected, which means $k = 1$ (see Figure 4). When the power switch Q is turned on, the currents flowing into C_1, C_2, C_3, and C_4 are I_{C1on}, I_{C2on}, I_{C3on}, and I_{C4on}; and the current flowing into L_a is I_{Laon}. I_{npon} is the current flowing into the primary side of the ideal transformer, while I_{nson} is the current flowing into the secondary side. Meanwhile, the current flowing into L_M is I_{LMon}, and the current flowing into L_r is I_{Lron}. I_{D2} is the current of D_2. I_{inon} and I_{outon} are the input current and the output current, respectively.

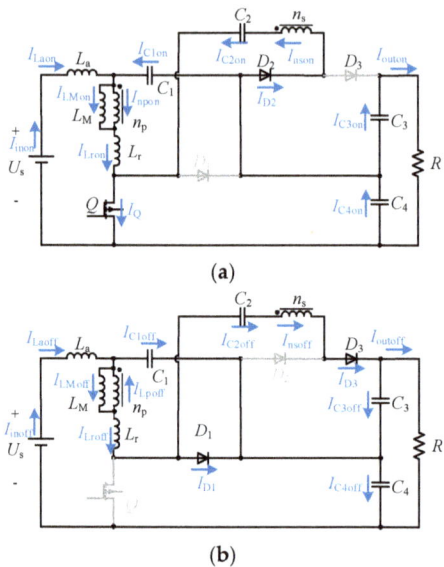

Figure 4. The current distribution in different states: (a) Q is turned on; (b) Q is turned off.

When the power switch Q is turned off, the currents flowing into C_1, C_2, C_3, and C_4 are I_{C1off}, I_{C2off}, I_{C3off}, and I_{C4off}; the current flowing into L_a is I_{Laoff}; and the currents flowing into both sides of the ideal transformer are I_{npoff} and I_{nsoff}. I_{LMoff} is the current flowing into L_M, and the current flowing into L_r is I_{Lroff}. I_{D1} is the current of D_1, while I_{D3} is the current of D_3. The input current and the output current are I_{inoff} and I_{outoff}, respectively.

By applying the ampere second balance principle to C_1–C_4, the equations can be derived as

$$\begin{cases} I_{C1on}d = I_{C1off}(1-d) \\ I_{C2on}d = I_{C2off}(1-d) \\ I_{C3on}d = I_{C3off}(1-d) \\ I_{C4on}d = I_{C4off}(1-d) \end{cases} \quad (14)$$

According to Figure 4a, it can be deduced that

$$\begin{cases} I_{C3on} = I_{outon} \\ I_{C1on} + I_{C2on} + I_{C3on} = I_{C4on} \\ I_{Laon} + I_{C1on} = I_{LMon} + I_{npon} \\ I_{npon} = nI_{nson} = nI_{C2on} \end{cases} \quad (15)$$

Similarly, Equation (16) can be derived from Figure 4b:

$$\begin{cases} I_{Laoff} = I_{C1off} + I_{LMoff} - I_{npoff} \\ I_{C2off} = I_{C3off} + I_{outoff} \\ I_{Laoff} - I_{outoff} = I_{C4off} \\ I_{npoff} = nI_{nsoff} = nI_{C2off} \end{cases} \quad (16)$$

Considering that the load is resistive, and the currents flowing into L_a and L_M are assumed to be constant, Equation (17) can be derived:

$$\begin{cases} I_{outon} = I_{outoff} = I_{out} \\ I_{Laon} = I_{Laoff} = I_{La} \\ I_{LMon} = I_{LMoff} = I_{LM} \end{cases} \quad (17)$$

In terms of Equations (14)–(17), the capacitor currents can be deduced as Equations (18) and (19):

$$\begin{cases} I_{C1on} = \frac{n}{d}I_{out} \\ I_{C1off} = \frac{n}{1-d}I_{out} \\ I_{C2on} = \frac{1}{d}I_{out} \\ I_{C2off} = \frac{1}{1-d}I_{out} \\ I_{C3on} = I_{out} \\ I_{C3off} = \frac{d}{1-d}I_{out} \\ I_{C4on} = \frac{1+n+d}{d}I_{out} \\ I_{C4off} = \frac{1+n+d}{1-d}I_{out} \end{cases} \quad (18)$$

$$\begin{cases} I_{La} = \frac{n+2}{1-d}I_{out} \\ I_{LM} = \frac{n+2}{1-d}I_{out} \end{cases} \quad (19)$$

The current stresses of the power switch Q and the diode D_2 can be obtained according to Figure 4a, Equations (18) and (19):

$$I_Q = I_{La} + I_{C4on} - I_{C3on} = \frac{n+2d}{(1-d)d}I_{out}, \quad (20)$$

$$I_{D2} = I_{C2on} = \frac{1}{d}I_{out}. \quad (21)$$

Similarly, the current stresses of D_1 and D_3 can be obtained by means of Figure 4b, Equations (18) and (19):

$$I_{D1} = I_{La} - I_{C1off} - I_{C2off} = \frac{1}{1-d}I_{out}, \quad (22)$$

$$I_{D3} = I_{C2off} = \frac{1}{1-d}I_{out}. \quad (23)$$

4.3. Performance Comparisons

The proposed converter was compared to the converter in Reference [29] without considering the influence of leakage inductance. The specific comparisons are shown in Table 1. The voltage stress across the diode D_1 was larger than the output voltage in the converter of Reference [29], while the voltage stresses across all semiconductors in the proposed converter were less than the output voltage. Moreover, the proposed converter could achieve a much higher voltage gain at the cost of one more diode and one more capacitor. The voltage gain M versus the duty cycle d is shown in Figure 5 when the turns ratio n is equal to 1 and 2.

Table 1. Comparisons of two types of converters with input current ripple minimization.

Characteristics	Converter in [29]	Proposed Converter
Voltage Gain	$(nd+1)/(1-d)$	$(n+2)/(1-d)$
Voltage Stress of power switch	$\frac{1}{1-d}U_{in}$	$\frac{1}{1-d}U_{in}$
Voltage Stress of diodes	$D_C: \frac{1}{1-d}U_{in}$ $D_1: \frac{n+1}{1-d}U_{in}$	$D_1: \frac{1}{1-d}U_{in}$ $D_2, D_3: \frac{n+1}{1-d}U_{in}$
Current Stress of power switch	$\frac{nd+1}{(1-d)d}I_{out}$	$\frac{n+2d}{(1-d)d}I_{out}$
Current Stress of diodes	$D_C: \frac{nd+1}{2(1-d)d}I_{out}$ $D_1: \frac{nd+1}{(1-d)d}I_{out}$	$D_1, D_3: \frac{1}{1-d}I_{out}$ $D_2: \frac{1}{d}I_{out}$
Number of power switches	1	1
Number of diodes	2	3
Ripple-minimization input current	Yes	Yes

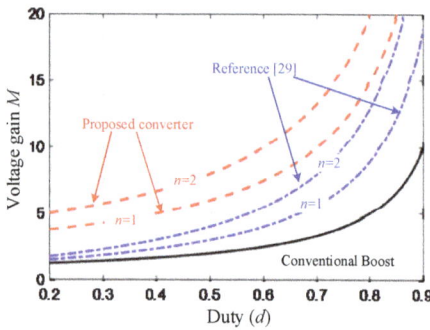

Figure 5. Curves of voltage gain versus duty cycle for two types of converters with input current ripple minimization.

5. Experimental Results and Analysis

In order to validate the feasibility of the proposed converter, a 400-W prototype was developed, as shown in Figure 6. An adjustable DC source with a voltage range of $U_{in} = 30$–100 V was used as the input of the converter to simulate a fuel cell source. The output voltage was controlled to 400 V by a voltage loop implemented on a digital signal processor (DSP) TMS320F28335. The experiment parameters are listed in Table 2. According to Equation (1), the current ripple coefficient α is defined as 0.2, and a large L_a can reduce the input current ripple: Therefore, L_a and L_M were taken as 241 µH and 368 µH, respectively. From Equation (2), the voltage ripple Δu is defined as 0.5 V. Considering practical experience and the laboratory conditions, C_2–C_4 were taken as 540 µF, and C_1 was taken as 270 µF.

The voltage stresses across all of the semiconductors for $d = 0.625$ and $U_{out} = 400V$ are shown in Figure 7. Figure 7a shows the voltage stresses across Q and D_1, which were equal to 133 V, while Figure 7b shows the voltage stresses across D_2 and D_3, which were 267 V, which was consistent with the theoretical analysis. In addition, the energy stored in the leakage inductance was released through the diode D_1 when Q was turned off. As a result, the voltage spike across the power switch Q was reduced significantly. The current flowing into the inductor L_a and the voltage across the capacitor C_1 are shown in Figure 8. The input current I_{in}, which flowed into L_a, had no fluctuation in each switching period, and neither did the voltage across C_1. Thus, it could be concluded that input current ripple minimization was achieved, as analyzed in Section 3.2. Figure 9 shows the currents flowing into both sides of the coupled inductors. The capacitor C_2 received energy from the coupled inductors when Q was turned on. On the contrary, the current I_{Lp} flowing into the primary side started to decrease when Q was turned off, C_2 started to discharge, and the current I_{Ls} flowing into the secondary side transferred energy to the load through D_3. The experimental results were consistent with the theoretical analysis.

Figure 6. The experimental prototype of the proposed converter.

Table 2. Experiment parameters.

Parameters	Values
Rated power P_n	400 W
Switching frequency f_s	20 kHz
Capacitor C_1	270 μF
Capacitors C_2, C_3, and C_4	540 μF
Inductor L_a	241 μH
Magnetizing inductor L_M	368 μH
Leakage inductance L_r	3.25 μH
Turns ratio $n_p:n_s$	1:1
Output voltage U_{out}	400 V
Input voltage U_{in}	30–100 V
Power switch Q	IXTH88N30P
Diodes D_1–D_3	DPG60C300HB

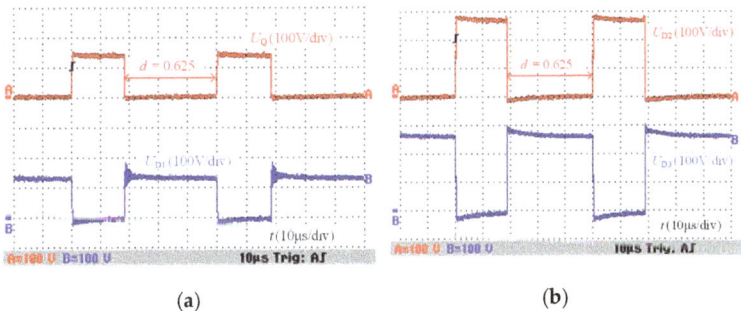

(a)　　　　　　　　　(b)

Figure 7. Voltage stresses across all semiconductors when U_{in} = 50 V and U_{out} = 400 V: (**a**) Voltage stresses across Q and D_1; (**b**) voltage stresses across D_2 and D_3.

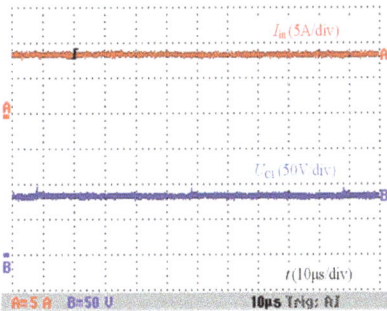

Figure 8. The current flowing into the inductor L_a and the voltage across the capacitor C_1.

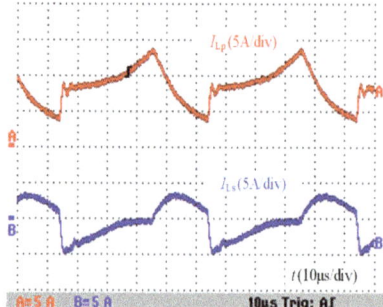

Figure 9. The currents flowing into both sides of the coupled inductor.

Figure 10 shows that with the voltage control loop, the output voltage U_{out} could be kept constant at 400 V, even when the input voltage U_{in} varied from 30 V to 100 V continuously. This demonstrated that the proposed converter could operate well in a wide voltage gain range from 13.33 to 4.

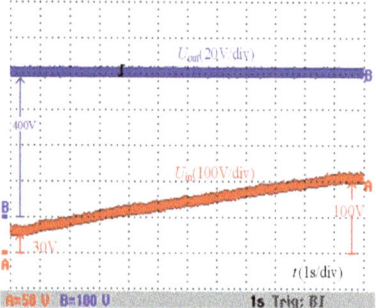

Figure 10. The output voltage and the wide-range input voltage from 30 V to 100 V.

The efficiencies of the proposed converter with different input voltages and different powers were measured by a power analyzer (YOKOGAWA/WT3000), as shown in Figure 11. When the input voltage U_{in} = 100 V and the load power P = 400 W, the proposed converter reached a maximum conversion efficiency of 95.12%. The minimum efficiency was 89.24% when the input voltage U_{in} = 40 V and the load power P = 400 W. To sum up, with the same power, as the voltage gain increased (i.e., the input voltage decreased), the efficiency of the proposed converter decreased. Since the input current rose as the low-side voltage decreased, the copper loss and the switching losses of the converter rose. Therefore, the efficiency of the converter appeared to degrade at a high voltage gain.

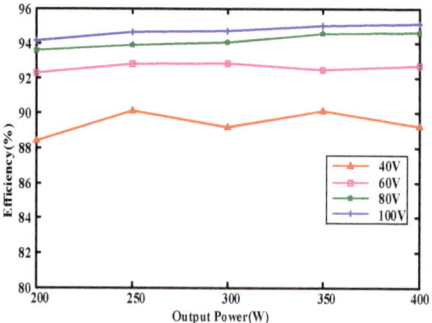

Figure 11. Efficiencies of the proposed converter with different output powers at different input voltages (i.e., different voltage gains) when U_{out} = 400 V.

6. Conclusions

A coupled-inductor DC-DC converter with a high voltage gain and input current ripple minimization was proposed in this paper. The converter can obtain a wide voltage gain range, and the voltage stresses across all semiconductors are lower than the output voltage. In addition, the proposed converter can benefit from the minimization of input current ripples. Furthermore, the passive lossless clamping circuit effectively suppresses voltage spikes caused by leakage inductance. Therefore, it is a good candidate for the power interface of fuel cell vehicles.

Author Contributions: Conceptualization, Y.Z.; methodology, J.L.; software, C.Z.; validation, C.D.; formal analysis, W.Z.; investigation, F.Y.; resources, F.Y.; data curation, C.Z.; writing—original draft preparation, C.Z. and F.Y.; writing—review and editing, J.L., and W.Z.; visualization, C.D.; supervision, Y.Z.; project administration, Y.Z.; funding acquisition, Y.Z.

Funding: This research was funded by Natural Science Foundation of China, grants number 51577130 and 51207104.

Conflicts of Interest: The authors declare no conflict of interest.

References

1. Khaligh, A.; Li, Z. Battery, ultracapacitor, fuel cell, and hybrid energy storage systems for electric, hybrid electric, fuel cell, and plug-in hybrid electric vehicles: State of the art. *IEEE Trans. Veh. Technol.* **2016**, *59*, 2806–2814. [CrossRef]
2. Bahroon, D.A.; Rahman, H.A. Performance evaluation of solar-fossil fuel-hybrid parabolic trough power plant in Yemen under different fuel types. In Proceedings of the 2014 IEEE Conference on Energy Conversion (CENCON), Johor Bahru, Malaysia, 13–14 October 2014; pp. 158–163.
3. Zeng, B.; Zhang, J.; Yang, X.; Wang, J.; Dong, J.; Zhang, Y. Integrated planning for transition to low-carbon distribution system with renewable energy generation and demand response. *IEEE Trans. Power Syst.* **2014**, *29*, 1153–1165. [CrossRef]
4. Boulanger, A.G.; Chu, A.C.; Maxx, S.; Waltz, D.L. Vehicle electri- fication: Status and issues. *Proc. IEEE.* **2011**, *99*, 1116–1138. [CrossRef]
5. Li, K.; Chen, T.; Luo, Y.; Wang, J. Intelligent environment-friendly vehicles: concept and case studies. *IEEE Trans. Intell. Transp. Syst.* **2011**, *13*, 318–328. [CrossRef]
6. Jin, J.X.; Chen, X.Y.; Wen, L.; Wang, S.C.; Xin, Y. Cryogenic power conversion for SMES application in a liquid hydrogen powered fuel cell electric vehicle. *IEEE Trans. Appl. Supercond.* **2015**, *25*, 1–11.
7. Su, W.; Rahimi-Eichi, H.; Zeng, W.; Chow, M.-Y. A survey on the electrification of transportation in a smart grid environment. *IEEE Trans Ind. Informat.* **2012**, *8*, 1–10. [CrossRef]
8. Jiang, W.; Fahimi, B. Active current sharing and source management in fuel cell-battery hybrid power system. *IEEE Trans. Ind. Electron.* **2010**, *57*, 752–761. [CrossRef]
9. Restrepo, C.; Konjedic, T.; Guarnizo, C.; Aviñó-Salvadó, O.; Calvente, J.; Romero, A.; Giral, R. Simplified mathematical model for calculating the oxygen excess ratio of a PEM fuel cell system in real-time applications. *IEEE Trans. Ind. Electron.* **2014**, *61*, 2816–2825. [CrossRef]
10. Rathore, A.K.; Bhat, A.K.S.; Oruganti, R. Analysis, design and experimental results of wide range ZVS active-clamped L-L type currentfed DC-DC converter for fuel cell to utility interface application. *IEEE Trans. Ind. Electron.* **2012**, *59*, 473–485. [CrossRef]
11. Jang, M.; Agelidis, V.G. A boost-inverter-based, battery-supported, fuel-cell sourced three-phase stand-alone power supply. *IEEE Trans. Power Electron.* **2014**, *29*, 6472–6480. [CrossRef]
12. Tseng, K.C.; Lin, J.T.; Huang, C.C. High step-up converter with three-winding coupled inductor for fuel cell energy source applications. *IEEE Trans. Power Electron.* **2015**, *30*, 574–581. [CrossRef]
13. Wang, P.; Zhao, C.; Zhang, Y.; Li, J.; Gao, Y. A bidirectional three-level dc-dc converter with a wide voltage conversion range for hybrid energy source electric vehicles. *J. Power Electron.* **2017**, *17*, 334–345. [CrossRef]
14. Qian, W.; Cao, D.; Cintron-Rivera, J.G.; Gebben, M.; Wey, D.; Peng, F.Z. A switched-capacitor DC-DC converter with high voltage gain and reduced component rating and count. *IEEE Trans. Ind. Appl.* **2012**, *48*, 1397–1406. [CrossRef]
15. Wu, B.; Li, S.; Smedley, K.; Singer, S. A family of two-switch boosting switched-capacitor converters. *IEEE Trans. Power Electron.* **2015**, *30*, 5413–5424. [CrossRef]

16. Wai, R.J.; Duan, R.Y. High step-up converter with coupled-inductor. *IEEE Trans. Power Electron.* **2005**, *20*, 1025–1035. [CrossRef]
17. Yang, L.S.; Liang, T.J.; Lee, H.C.; Chen, J.F. Novel high step-up DC-DC converter with coupled-inductor and voltage-doubler circuits. *IEEE Trans. Ind. Electron.* **2011**, *58*, 4196–4206. [CrossRef]
18. Ajami, A.; Ardi, H.; Farakhor, A. A novel high step-up DC/DC converter based on integrating coupled inductor and switched-capacitor techniques for renewable energy applications. *IEEE Trans. Power Electron.* **2015**, *30*, 4255–4263. [CrossRef]
19. Dwari, S.; Parsa, L. An efficient high-step-up interleaved DC-DC converter with a common active clamp. *IEEE Trans. Power Electron.* **2011**, *26*, 66–78. [CrossRef]
20. Zhao, Y.; Li, W.; Deng, Y.; He, X. High step-up boost converter with passive lossless clamp circuit for non-isolated high step-up applications. *IET Trans. Power Electron.* **2011**, *4*, 851–859. [CrossRef]
21. Thounthong, P.; Sethakul, P.; Davat, B. Modified 4-phase interleaved fuel cell converter for high-power high-voltage applications. In Proceedings of the 2009 IEEE International Conference on Industrial Technology, Gippsland, VIC, Australia, 10–13 February 2009.
22. Hu, X.; Gong, C. A high gain input-parallel output-series DC/DC converter with dual coupled inductors. *IEEE Trans. Power Electron.* **2015**, *30*, 1306–1317. [CrossRef]
23. Tan, S.-C.; Kiratipongvoot, S.; Bronstein, S.; Ioinovici, A.; Lai, Y.M.; Tse, C.K. Adaptive mixed on-time and switching frequency control of a system of interleaved switched-capacitor converters. *IEEE Trans. Power Electron.* **2011**, *26*, 364–380. [CrossRef]
24. Tseng, K.-C.; Huang, C.-C. High step-up high-efficiency interleaved converter with voltage multiplier module for renewable energy system. *IEEE Trans. Ind. Electron.* **2014**, *61*, 1311–1319. [CrossRef]
25. Zhan, Y.; Guo, Y.; Zhu, J.; Li, L. Performance comparison of input current ripple reduction methods in UPS applications with hybrid PEM fuel cell/supercapacitor power sources. *Int. J. Electr. Power Energy Syst.* **2015**, *64*, 96–103. [CrossRef]
26. Zhan, Y.; Guo, Y.; Zhu, J.; Li, L. Input current ripple reduction and high efficiency for PEM fuel cell power conditioning system. In Proceedings of the 2017 20th International Conference on Electrical Machines and Systems (ICEMS), Sydney, NSW, Australia, 11–14 August 2017.
27. Paja, C.A.R.; Arango, E.; Giral, R.; Montesa, A.J.S.; Carrejo, C. DC/DC pre-regulator for input current ripple reduction and efficiency improvement. *Elect. Power Syst. Res.* **2011**, *81*, 2048–2055. [CrossRef]
28. Shi, Y.; Liu, B.; Duan, S. Low-frequency input current ripple reduction based on load current feedforward in a two-stage single-phase inverter. *IEEE Trans. Power Electron.* **2016**, *31*, 7972–7985. [CrossRef]
29. Chen, Z.; Xu, J. High boost ratio DC-DC converter with ripple-free input current. *Electron. Lett.* **2014**, *50*, 353–355. [CrossRef]

© 2019 by the authors. Licensee MDPI, Basel, Switzerland. This article is an open access article distributed under the terms and conditions of the Creative Commons Attribution (CC BY) license (http://creativecommons.org/licenses/by/4.0/).

Review

Electric Vehicles for Public Transportation in Power Systems: A Review of Methodologies

Jean-Michel Clairand [1,*], Paulo Guerra-Terán [1], Xavier Serrano-Guerrero [2,3], Mario González-Rodríguez [1,4] and Guillermo Escrivá-Escrivá [3]

1. Facultad de Ingeniería y Ciencias Agropecuarias, Universidad de las Américas—Ecuador, Quito 170122, Ecuador
2. Grupo de Investigación en Energías, Universidad Politécnica Salesiana, Cuenca 010103, Ecuador
3. Institute for Energy Engineering, Universitat Politècnica de València, 46022 Valencia, Spain
4. Intelligent & Interactive Systems Lab (SI² Lab), Universidad de las Américas—Ecuador, Quito 170125, Ecuador
* Correspondence: jean.clairand@udla.edu.ec; Tel.: +593-9-95860613

Received: 3 July 2019; Accepted: 9 August 2019; Published: 14 August 2019

Abstract: The market for electric vehicles (EVs) has grown with each year, and EVs are considered to be a proper solution for the mitigation of urban pollution. So far, not much attention has been devoted to the use of EVs for public transportation, such as taxis and buses. However, a massive introduction of electric taxis (ETs) and electric buses (EBs) could generate issues in the grid. The challenges are different from those of private EVs, as their required load is much higher and the related time constraints must be considered with much more attention. These issues have begun to be studied within the last few years. This paper presents a review of the different approaches that have been proposed by various authors, to mitigate the impact of EBs and ETs on the future smart grid. Furthermore, some projects with regard to the integration of ETs and EBs around the world are presented. Some guidelines for future works are also proposed.

Keywords: charging approaches; electric bus; electric taxi; electric vehicle; public transportation; smart grid

1. Introduction

Transportation is one of the sectors that is facing various challenges due to environmental concerns. These concerns include the depletion of fossil fuels, global warming, and local pollution. In this scenario, battery-powered electric vehicles (EVs) could be a proper solution to mitigate environmental issues [1,2]. These are wheeled vehicles that use an electric motor that is powered by a battery for propulsion. They must be recharged at home or at a public charging station, where three main levels of charging are used, depending mainly on the charge rate [3]. For the purposes of this paper, an EV will be a vehicle supplied with electricity from a battery, in order to differentiate from electric trains or trolleybuses, which are vehicles supplied with electricity directly from the grid by means of overhead wires.

In particular, EVs do not generate local pollution, and they have a well-to-wheel energy efficiency that is much more significant than that of internal combustion vehicles (ICVs). However, EVs present cradle-to-grave environmental impacts, especially due to the use of lithium batteries. The manufacturing phase corresponds to the highest environmental burden of EVs, mainly in the toxicity categories because of the use of metals in the battery pack [4]. To address these issues, it is crucial to minimize power losses in the battery and develop proper recycling tools [5]. Despite these issues, EVs can reduce CO_2 emissions with most of the generation mix scenarios. If the electricity is generated only by coal plants, the well-to-wheel CO_2 emissions of EVs are still similar to that of ICVs [6–8]. Therefore, several governments in different countries are promoting the purchase of EVs with economic incentives. The replacement of ICVs with EVs will offer the potential to significantly reduce greenhouse gas emissions.

Although EVs present several environmental advantages, a massive introduction of them could create several issues in the power grid, which has been studied by several researchers. For example, in [9], the impact of different penetration levels of plug-in EVs in a distribution system was considered, and it was demonstrated that a significant EV load leads to voltage drop and voltage deviations. A 30% plug-in EV integration in the grid led to a voltage deviation of 10.3% between 18 h00–21 h00 in the winter period, in the studied case. In [10], it was demonstrated that charging EVs considerably increases the distribution load and, so, the total power losses. Furthermore, EV charging increases the daily peak load. The authors of [11] indicated that EVs generate substantial investment costs in distribution systems, and that power losses can reach up to 40% for an EV penetration of 62%. In [12], it was exposed that EV fast charging leads to harmonic issues and failure to respect IEEE standard limits. In [13], it was proposed that the life durations of low-voltage transformers are reduced with a high penetration of EVs.

Considering all these issues, several researchers have widely studied strategies for the massive introduction of EVs into power systems. For example, smart charging of EVs is an important area of study, which allows EV users and grid operators to properly manage EV charging profiles in order to obtain technical and economic benefits, as well as considering the specific demand-side management of EVs [11]. Smart charging techniques include the Vehicle-to-Grid (V2G) concept, where the EV not only charges from the grid, but it supplies energy when necessary, becoming a generation/storage device [14,15]. For instance, investments in Renewable Energy Sources (RESs) for cleaner electricity production have also been related to EV integration into the grid. In particular, RES such as solar photovoltaic (PV) and wind create other challenges, due to their power generation uncertainties and fluctuations, as well as their high installation costs [16,17].

Some other researchers have focused on the management of EV charging stations. In particular, it is crucial to locate the optimal placement of EV charging stations to meet technical grid constraints, considering customer wait times [18,19]. Another solution to mitigate EV impacts on the grid considers implementing smart chargers. These chargers are generally implemented off-board the EV and are connected by DC plug. They possess advanced communication capabilities to receive instructions from the grid operator, in order to take actions for the grid requirements. They are usually bidirectional to also supply electricity to the grid, enabling control options for distribution feeders [20–22].

On this basis, EV integration in the grid has widely been addressed in the literature. Various reviews have studied smart charging techniques, potentials, barriers, and technologies for EVs. For example, in [23], smart charging approaches, such as strategies, algorithms, methods, and projects, were presented. The authors of [24] reviewed the EV economic, environmental, and grid impacts, and the interactions between EVs and RESs. In [25], the main technical challenges for the integration of EVs into the grid were studied. The authors of [26–28] centered their review of V2G impacts, potentials, and limitations on RES integration with EVs. The challenges and opportunities for a Lithuanian case study were provided in [29]. In [30], optimization techniques for EV charging infrastructures, such as computational and algorithmic aspects, were analyzed. EV charging control and operation in power grids was studied in [31], considering issues related to the real-time EV charging in smart grids. Although EV integration into power systems has been appropriately studied in the literature, all these works (and others) have only addressed the issues relating to private EVs. To the knowledge of the authors, no work has yet reviewed the integration techniques in power systems for public transportation using EVs, which is the main objective of this work. In addition, it is necessary to understand the importance of the behavior of these new users, as their behavior in terms of schedules and demanded power is very different than that of private EVs.

So far, there has been scarce research in the literature on EVs for public transportation, such as taxis and buses. In particular, public transportation is crucial for modern societies with growing populations. Public transportation was defined as "a service provided by public or private agencies that is available to all persons who pay the prescribed fare" in [32]. Moreover, public transport has the following characteristics:

- Efficient mass transfer of passengers;
- Ease of access, such that any individual has the means to use public transport;
- Transport along an organized system of fixed routes; and
- Based on a pre-defined timetable, resulting in fixed intervals of transport along particular routes.

Generally, taxis meet the four conditions but, in several places, the first and third conditions are not met. Thus, the question of whether taxis are public transportation arises. However, this question is out of the topic, so ETs will be assumed as public transportation in this paper. Despite the environmental benefits of personal EVs, they may result in high congestion in urban areas, which creates another concern. Furthermore, even if EVs emit less pollutants than ICVs, they have a considerable footprint due to the required electricity generation. Investing in public transportation will considerably decrease CO_2 emissions [33,34], and will be more beneficial for the environment if the vehicles used are EVs. On the other hand, public transportation should also be electric for health reasons. It has been demonstrated that emissions from ICVs, especially from buses, create several health problems [35]. These problems have been shown to be worse, in terms of emissions, for high-elevation cities [36]. Therefore, central and local governments are pushing to promote EVs for public transportation. In terms of the aforementioned issue, it is crucial to improve the quality of the services offered by public transportation [37].

The challenges of using EVs for public transportation are much different from those of private EVs. Taxis travel much longer daily distances than an average private driver. Therefore, the daily energy needed to charge an electric taxi (ET) is much higher than a typical EV. Furthermore, as mentioned above, it is arguable that ETs are public transportation, because they do not have fixed routes and timetables, and these conditions must be taken into consideration. Electric buses (EBs) require high energy capacity batteries and, so, they will consume a significant amount of power during charging time, which will create an impact on the grid. These new issues must be adequately addressed by researchers, as well as transportation and energy players.

The aim of this paper is to review the different proposed approaches and tools used by researchers to study the impact of ETs and EBs in power systems. With this review, it is expected that researchers who are working on this topic could comprehend the state of the art and provide some insights into research gaps for future works.

The rest of this paper is organized as follows: Section 2 presents an overview of EVs for public transportation. Section 3 discusses the main approaches of integrating ETs and EBs into power systems. Section 4 presents some projects related to EV integration into public transportation. Section 5 provides some insights for future work. Finally, Section 6 highlights the main conclusions of the paper.

2. Electric Vehicles for Public Transportation

2.1. Types of EVs

As mentioned above, battery EVs are vehicles that are fully powered by a battery, in order to differentiate them from ICVs, hybrid EVs, and plug-in hybrid EVs. Typically, their distance range is much smaller than other kinds of vehicles, but they are much more efficient and environmentally friendly.

EVs are powered by electricity, which is stored in the battery. In addition, the configuration for both buses and sedans (for ETs) include auxiliary devices, an electric motor, a transmission system, and a final drive, as shown in Figure 1.

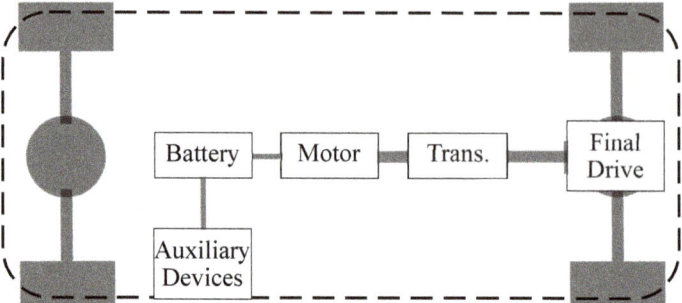

Figure 1. Electric Vehicle (EV) Configuration [38].

For private EVs, various models are sold in the market, such as the Nissan Leaf and Kia Soul EVs. However, these models may be not appropriate for taxis. In particular, the Nissan Leaf and Kia Soul have low distance ranges: 241 km for the Nissan Leaf and 179 km for the Kia Soul [39,40]. The most appropriate models for ETs considered by investors and researchers are the BYD e6 and the Tesla S [41,42], due to their higher battery capacities and distance ranges. The characteristics of these vehicles are summarized in Table 1.

For EBs, the brand BYD has produced various models and is the brand that has sold the most buses in the world [43]. In Europe, the most commercialized EBs are from Volvo and Solaris. Some of these models are, also, summarized in Table 1.

Table 1. Characteristics of EVs for public transportation [41–45].

Model	Use	Battery Capacity [kWh]	Charging Power [kW]	Range [km]	Charging Time [min]
BYD e6	Taxi	80	40	400	120
Tesla S	Taxi	100	16.5/120	480	420/42
BYD 23' Coach	Bus	121	40	200	180
BYD 40' Coach	Bus	352	40×2	322	270
BYD 60' Coach	Bus	578	200	355	180
Solaris Urbino 12	Bus	145	2×125	100	24
Volvo 7900	Bus	76	300	200	15

2.2. Charging EVs for Public Transportation

Charging systems are crucial components for the adoption of EVs in transportation. Charging EVs for public transportation is a key challenge, as mentioned above, because of the higher amount of required electrical energy for charging EV batteries, as well as the time requirements.

For ETs and EBs, three principal types of charging emerge: Plug-in charging, battery swapping, and wireless charging. These kinds of charging might be implemented in public charging stations, and must interact with power systems, the electricity market, and fleet operators, in order to meet their respective constraints. Power system constraints include power, voltage, and frequency limits. The electricity market provides electricity prices to optimize charging costs or, in some cases, electricity bids to participate in ancillary services. Fleet operator constraints include the schedules for EBs and ETs, which must be respected to guarantee passenger satisfaction. This interaction is depicted in Figure 2.

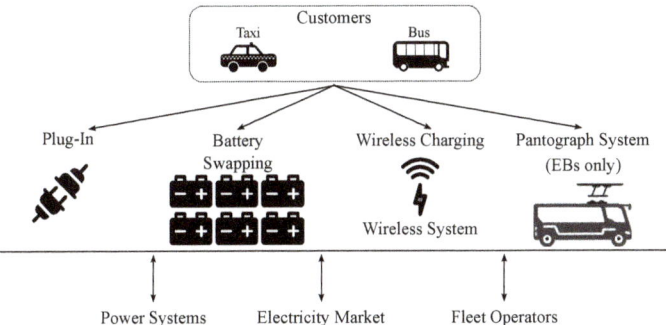

Figure 2. Interaction between EVs for public transportation and power systems.

2.2.1. Plug-In Charging

Plug-in charging corresponds to the most widely used method for charging EVs, especially for private ones. For small vehicles, such as ETs, three primary charging levels are used: Level 1, or slow charging, is the slowest charging, which is not recommended for EVs for public transportation, because of the time limitations. Level 2 requires a 240 V outlet and is the primary method for both public and private facilities. Level 3 requires a DC connection, and is the fastest solution for charging an EV battery, but requires a high amount of power over a small time duration [3]. These three levels are summarized in Table 2.

Table 2. Charging Power Levels [3].

Power Level Types	Charger Location	Expected Power Level	Charging Time
Level 1	On-board, 1-phase	1.9 kW (20 A)	11–36 h
Level 2	On-board, 1- or 3-phase	8 kW (32 A), 19.2 kW (80 A)	2–6 h
Level 3	Off-board, 3-phase	50 kW, 100 kW	0.2–1 h

For EBs, it is not feasible to consider slow charging levels, as the battery capacity (around 400 kWh) is high and, so, charging could take very long, limiting trip schedules. Fast charging seems the best level option for both EBs and ETs, considering consumer convenience. However, this charging level presents issues: EV batteries heat up while being charged, large conductors and AC/DC converters are required, and electric utilities limit their use because the critical power required creates grid issues. Hence, these technical limitations have been the focus of study of some works [46]. Some of the alternatives consist of providing energy storage by distributed generation in a fast-charging station to mitigate peak loads on the grid.

2.2.2. Wireless Charging System (WCS)

In addition to the issues presented before, environmental conditions (e.g., rain, snow, and extreme temperatures) can cause discomfort for users when they connect an EV manually at the charging station. Moreover, failure of power cords and connectors can cause safety issues (risk of electrical sparking and electrical shock) [47]. Wireless charging systems (WCS) arise as a solution for the above problems. Furthermore, the main advantage of WCS is providing the opportunity for making fast and frequent charges while the EV is in transit; for example, in streets with heavy traffic, bus stops, parking lots, and so on. This possibility allows for battery downsizing with the following main benefits: (a) lighter weight EVs, (b) reduction of cost of the EV and repositioned batteries, (c) savings in energy due to a decrease in mass, and (d) reduction of CO_2 emissions. For example, in [48], the authors concluded that a WCS, in comparison with a plug-in charging system, consumes 0.3% less energy and emits 0.5% fewer greenhouse gases. The WCS is a proven technology that provides an exciting alternative charging system, which has been used in several projects for city battery-powered buses [47]. The principal difference between a plug-in charging system and a WCS is that the second uses a pair of coupled coils or capacitors, instead of a transformer. The air-gap between the two coils or capacitors produces a leakage magnetic field that needs to be controlled for safety reasons [49]. Wireless charging has the following steps: First, a rectifier that converts the AC utility power to DC power; then, the DC power is converted to AC power of high frequency to drive the primary coil or capacitor by a compensation network. The electromagnetic field produced by the primary coil or capacitor induces an alternating voltage in the secondary coil or capacitor, transferring AC power that is then is rectified to charge the battery [49]. A WCS is depicted in Figure 3.

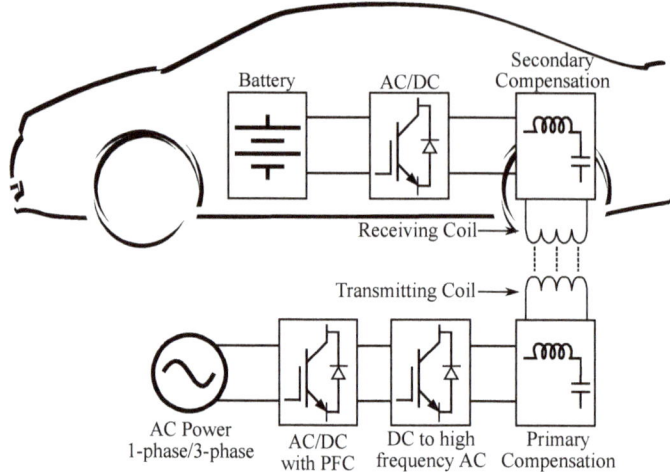

Figure 3. Wireless Charging system [49].

The literature shows that there are three main classifications of WCS that can be used for public buses: Capacitive wireless power transfer, inductive wireless power transfer, and resonant inductive power transfer [50,51].

The main advantages of WCSs include less energy consumption and less greenhouse gases emitted, compared to plug-in charging [48]. The challenges of this technology include low efficiency of energy transfer, due to losses during coil-to-coil transfer; high installation costs, which are significantly higher than plug-in charging; and human exposure to radio frequency radiation and magnetic fields [50].

2.2.3. Battery Swapping Station (BSS)

A Battery Swapping Station (BSS) is an EV station where customers can swap their discharged battery with a charged one [52]. It has begun to be adopted, especially for electrified public transportation. The BSS requires a high stock of batteries to supply its customers. Many works assume that the batteries are owned by the BSS and rented to the customers, which could be EB and ET companies. The main benefit of BSS for customers is that they can immediately have a charged battery, similar to gas stations. Moreover, the BSS generally provide benefits for the EV batteries, since their life is not affected when they are not charged at fast charging levels. This is possible when the amount of batteries owned by a BSS is high and, so, they have time flexibility to charge them.

The BSS commonly includes a vehicle platform, lift, alignment, and equipment rollers; battery lifts, conveyor shuttles, storage racks, and rails; and electrical connection alignments [53].

The electrical components of the BSS to charge the batteries include a distribution transformer, AC/DC chargers, battery packs, and a battery energy control module [53]. The transformer converts high voltage levels from the grid to a lower level, adapted to supply electricity to the batteries.

Then, as the batteries require DC energy, the AC/DC adapts the AC energy coming from the transformer. The control module allows charging at different power levels, depending on grid requirements. Some works have considered a bidirectional AC/DC converter, allowing V2G services.

One of the main advantages of BSSs is that a third party could own the batteries and be responsible for replacing them with fully charged ones, monitoring their health and decommissioning the batteries once they are no longer suitable. Moreover, the BSS offers time benefits to users, similar to typical gas stations, avoiding long wait times. The main limitations of BSS include standardization of EV battery packs, acceptance of the BSS model, and the reliable estimation of battery state-of-health. Furthermore, the question of whether BSS are profitable has been brought up in [53].

2.2.4. Pantograph System

A pantograph system is a solution that allows the EBs to charge quickly at stops, which has been the focus of various research studies. The pantograph system includes an automatic connecting system, DC-conductive charging supply equipment, fixed conductive rails attached to the roof of the vehicle, conductive poles, and communication systems [54]. The Automatic Connecting System controls and monitors a connection device for conductive charging fixed to the infrastructure above the vehicle. It includes an Automatic Connecting Device, which connects or disconnects EV supply conductive components to the vehicle interface. The DC EV conductive charging system provides a voltage in a range between 450–750 V, allowing a power supply up to 450 kW. A very well-known pantograph is the OPPCharge from ABB [55]. Figure 4 illustrates a pantograph system.

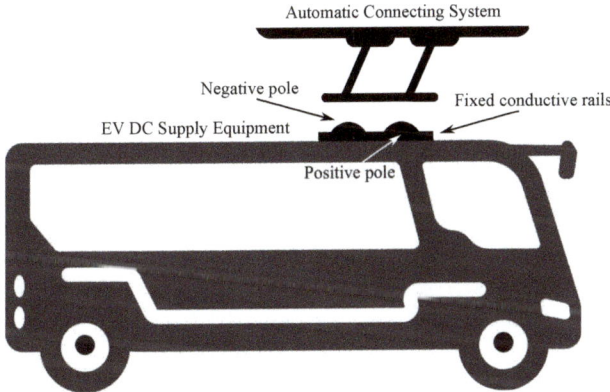

Figure 4. Pantograph System.

Even if several EB makers provide pantograph interfaces, BYD does not offer any pantograph system for their customers because their EBs have high distance range autonomy and can operate for an entire day, so only one daily charge for a few hours is required [56]. Moreover, so far, no research work has included pantographs in their approach.

3. Impact of Electric Vehicles for Public Transportation in Power Systems

The various approaches considering the introduction of EBs and ETs in Power Systems is a recent trend that has been studied by several researchers. Note that research in this field only began in 2014. So far, the research in this area has been much smaller than that on the impact of private EVs on power systems. A search in SCOPUS was performed, finding separately the approaches for ETs, EBs, and the few approaches that considered both ETs and EBs. Therefore, all the research papers related to this topic were analyzed and are summarized here. In Figure 5, the number of peer-reviewed papers published by some countries on EB and ET integration into the grid is shown. It can be seen that China leads by far, in terms of paper production, followed by the USA. This can be explained by the fact that several real projects using ETs and EBs are being developed in various cities in China.

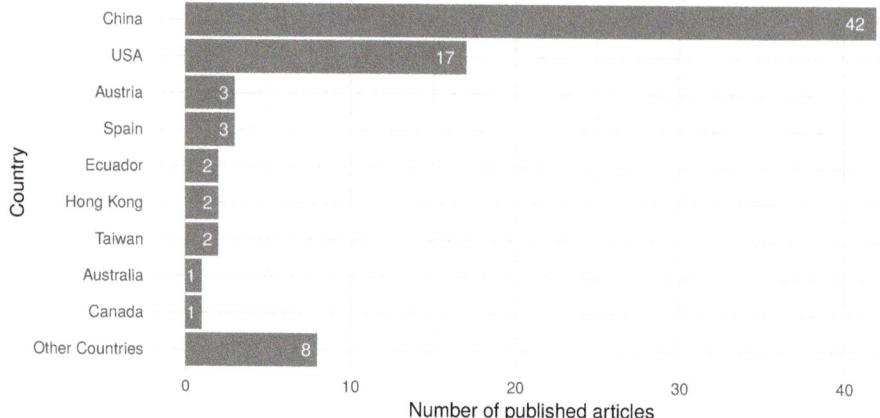

Figure 5. Number of published articles from different countries.

3.1. Electric Taxis (ETs) Approaches

In the literature, various methodologies for integrating electric taxis (ETs) into the grid can be found. Foremost, several researchers have studied the load prediction of ETs, in order to evaluate the impact on the grid. Thus, the authors of [57] proposed a precise charging load model for ETs based on multi-step Q(λ) learning. In [58], the GPS measurements of 460 San Francisco taxis were used to estimate the charging load. Furthermore, an economic analysis was performed, demonstrating that an ET fleet would reduce the costs considerably, compared to an ICV fleet. In [59], the charging loads of ETs were studied, based on real data from the taxi charging data of 129 vehicles from Shenzhen, China. In [60], the distribution of ET charging demands were studied based on resident's travel characteristics. The authors of [61] studied the feasibility of a purely ET fleet, based on the ZENEM demonstration project in Vienna, Austria. The results demonstrated that the main issue of an ET fleet connected to the grid results in overloading, which needs to be appropriately managed by controlled charging. In [62], battery charging and swapping stations were modeled to compare their load impacts, demonstrating that EV fast charging generates more important peaks than swapping stations. The authors of [63] performed a realistic large-scale ET application in Beijing, based on the real in-use EV data to analyze fleet driver behavior, charging patterns, and energy consumption and efficiency. Note that the results were representative for this particular case study.

Some other works put focus on the charging station sites and sizing for ETs. Thus, in [64], a decision support system was studied for placing ET charging stations. The objective was to maximize charging demand satisfaction of ET drivers, based on the data of 800 taxi vehicles in Vienna. The authors of [65] exposed a data-driven optimization-based approach for siting and sizing ET charging stations. In this case, the objective was to minimize the charging infrastructure by using integer linear programming. The case study corresponded to Changsha, China, where data was obtained from the large-scale GPS trajectory of the taxi fleet. The results indicated that various candidate sites did not need to install chargers and that, allowing for a waiting time, the numbers of chargers can be significantly reduced. However, the authors do not account for the SOC when ETs arrive at charging station. Similarly, the optimal location of charging stations was presented in [66]. Their methodology was based on a multi-agent system to simulate plug-in hybrid ET daily operations, and minimized charging costs of plug-in ETs, power losses, and voltage deviations. The novelty of their approach was that taxi agents made decisions concerning whether to find passengers or to charge the plug-in ETs. Furthermore, this model included the participation of new agents for ETs, such as the time agent, the map agent, the passenger generator agent, the charging station agent, and plug-in ET agent. However, this methodology included a reward function that depended on the time for finding passengers, which could be feasible for this particular case study, but the behavior of ETs in the real world is significantly different and depends mostly on local laws. The authors of [67] considered the deployment of charging stations, considering the effects of passengers, taxi drivers, electricity retailers, the transportation and distribution networks, and power consumers. To solve the proposed model, a multi-objective optimization was proposed. In [68], the siting and sizing were determined based on the calculation of the radius of the charging station service.

Some other approaches considered the optimal ET plug-in charging station time usage. Thus, in [69], a multi-objective optimization was proposed to maximize the use efficiency of charging facilities, minimize the load unbalance (power losses), and minimize the customer cost, by use of a fuzzy mathematical method and improved particle swarm optimization algorithm. Data of Shenzhen was used to demonstrate the effectiveness of the methodology. The authors of [70] proposed a methodology that assists ET to charge more efficiently with respect to time. Actual data from eight charging stations in Shenzhen was used to simulate the proposed model. However, no conditions for the grid were considered.

Some attention has been paid to system costs. In [71], the charging costs were defined as the loss of service income caused by charging. A plug-in ET driver can choose the ideal time slot to charge, considering the SOC, time-varying income, and queuing status at charging stations. Cost minimization was obtained through a game-theoretic approach, and this method demonstrated that it is also possible to enhance the use ratio for charging stations, and to flatten the unevenness of charging request for power grids. The authors of [72] studied the interaction between a fleet of self-driving EVs, which can be considered to be ETs, and the grid. A joint model that captured the coupling between the EV charging requirements, capturing time-varying customer demand, battery depreciation, and grid transmission constraints, was considered. A Texas case study was proposed, demonstrating the effectiveness of the model. The results indicated that system costs were reduced.

Focusing on BSSs, it is crucial to develop methodologies to maximize the daily operation profits. Thus, the authors of [73] proposed a linear programming model by considering constraints on the battery swapping demand of users and charging state of batteries. A real case study from the ETs in Beijing was used to prove the effectiveness of the methodology.

Some approaches have focused on minimizing the costs of BSSs. For example, the authors of [74] presented a Smart Battery Charging and Swapping Operation Service Network for both particular EVs and ETs. The methodology was based on the State Grid Corporation of China, which consists of providing battery swapping as the primary service (complemented by typical charging, if necessary). A model based on Queuing theory was proposed to minimize the overall cost of station operations, considering that most of the taxis charge 3 or 4 times per day, with a fast charging time of 46 min. A case study in Hangzhou, China, was proposed to demonstrate the effectiveness of the methodology. Another example was performed by [75]. A Monte Carlo method and the Dijkstra Algorithm were adopted to simulate electric taxi operations. Then, an optimization model was proposed to minimize the costs of the BSS. In [76], a real-time battery scheduling problem was proposed. The strategy was evaluated using the data set of routes in the city of Suzhou, China. In the same line, the authors of [77] proposed a methodology that minimizes electricity costs, battery degradation, and low battery use. The problem was solved by convex optimization with both spatially and temporally coupled constraints.

Real-Time pricing has also been adopted as a methodology. In [78], the profit maximization of a plug-in ET was investigated, considering the uncertainties of electricity prices and time-varying incomes, which was solved by a thresholding method. The results demonstrated improvements in profits compared to common strategies. The authors of [79] studied the optimal charging problem to maximize the average profit of a plug-in ET in the long-term, considering SOC dynamics constraints. The optimization problem is infinite binary and was solved by dividing it into a series of periodic sub-problems. The proposed method demonstrated better performance than other algorithms, such as stochastic strategy. In [80], real-time pricing was proposed for regulating the collective charging load of the ET fleet. This could induce the taxi fleet to modify its charging decisions. However, the charging depended on taxi drivers and, so, the fluctuation of charging prices could not be used and the high prices for charging in peak periods could not be ignored.

The massive introduction of ETs also creates economic challenges. Therefore, the cost-effectiveness of investments must be studied. In particular, the question concerning charging or battery changing for ETs arises. Hence, in [81], the long-term planning for ET introduction has been studied by comparing battery swapping and battery charging models in the Pinggu District of Beijing. The results indicated that for this particular case study, the battery swapping model was economically better than the battery charging model, but a high level of technology must be considered. The authors of [82] proposed a MILP optimization to minimize the investments and operation costs of ET charging stations, considering the constraints of the service radius of charging stations, charging demand satisfaction, and rational occupation rate of chargers.

Another novelty concerning typical EVs is the use of taxi apps. Taxi apps allow for the finding of customers, which reduces empty cruising distance. It has also been considered that taxi apps will enable drivers to find the closest charging stations, with real-time information about charging availability. The authors of [83] demonstrated that taxi apps reduce taxi trip distances and, therefore that they can reduce energy consumption for EVs.

The analyzed ET strategies are summarized in Table 3, and the tools used in Table 4.

Table 3. Main Objectives of ET-related Papers.

Main Objectives	References
ET Load prediction	[57–63]
Siting and sizing of ET charging stations	[64–68]
Optimal plug-in charging station time usage	[69,70]
Minimized Costs of ET plug-in charging stations	[71,72]
Maximization of Profit of BSS	[73]
Minimized Costs of BSSs	[74–77]
Minimized Load Unbalance	[69]
Minimized Customer Cost	[69]
Real-Time Pricing (decentralized)	[78–80]
Planning ET charging stations	[81,82]
Taxi Apps	[83]

Table 4. Tools used.

Tools Used	References
Method	
Integer Linear Programming	[65]
Linear Programming	[73]
Multi-Objective Optimization	[66,67,69]
MILP Optimization	[64,82]
Convex Optimization	[77]
Particle Swarm Optimization	[69,70]
Binary programming	[79]
Game-theoretic approach	[71]
Heuristic Method	[76,84]

3.2. Electric Bus (EB) Approaches

Several researchers have studied the load prediction of EBs. For example, in [85], the forecasting of EB BSS was proposed, based on stochastic modeling. The charging load model used statistical data of the travel patterns of EBs. The hourly number of EBs, starting charging time, travel distance, and charging duration were considered to be crucial variables for the forecast and were modeled through prediction methods, such as neural networks, uniform distributions, and the Gaussian model. A Monte Carlo method and a kernel density estimator were used for handling the uncertainties. The authors of [86] studied the modeling of EB buses in a full transit network, based on a real-time simulation model considering the real-world transit constraints of Belleville, Ontario, Canada. Moreover, the impact of the grid has been analyzed, such as the impacts on the lifetimes of substation transformers and the voltage regulation and voltage control devices. The results indicated that EB loads reduce transformer lifetimes considerably and generate voltage issues. In [87], short-term forecasting was studied for EB charging stations. This forecasting was based on a hybrid model, which combines fuzzy clustering, a least squares support vector machine, and the wolf pack algorithm (WPA). To demonstrate the accuracy of the forecasting, two case studies were proposed, demonstrating high precision.

Some approaches have focused on minimizing the charging costs of EB fast-charging stations. In [88], a methodology that reduced the total costs of investments and charging costs was presented. This work considered the value of energy storage in an EB fast-charging station, demonstrating that energy storage contributes to the reduction of long-term costs. The problem was solved by a mixed integer non-linear programming formulation, considering the capital costs of the transformer, distribution feeder, and energy storage constraints. The authors of [89] proposed a charging strategy for fast-charging stations based on a decision-making process, which considered that the EBs only charge when the SOC is below a charging threshold at the fast charger. This strategy was simulated using a case study of Tallahassee, Florida, demonstrating cost reductions compared to cases without charging strategies. In [90], real-time coordinated charging strategies for EB fast-charging stations were proposed. The purpose of this was to minimize power purchase costs, considering the daily Time-of-Use prices, and the peak loads were also mitigated. This work was complemented by [91], where an additional energy storage system was included, and the problem formulation was more complete. A heuristic method was used, considering whether the plug-in EBs were controllable or not. The method was simulated in Chongqing, China, validating the importance of the energy storage in EB charging stations.

Minimizing costs of EB BSSs has also been considered by some authors. In [92], the operational costs of an EB BSS were minimized. For this, the EB load was forecasted, based on a fuzzy evaluation and through actual survey data. The charging costs were optimized based on a Genetic Algorithm optimization approach and considering Time-of-Use prices. The case study of Baoding, Hebei province, China was evaluated, indicating an improvement in the profits with normal conditions. In the same line, the authors of [93] minimized the cost of a BSS with distributed PV.

Focusing on charging stations, it is essential to study their allocation. In this aspect, the deployment of EB charging stations has been considered in [94]. A bi-objective optimization was used to minimize both the number of charging stations and the EB stop time. For this, based on a discrete event simulation, the EB energy consumption was evaluated, taking into account load and friction forces, as well as different data from a case study of Curitiba, Brazil, such as passenger demand, bus, speed, distances, and route elevation profiles. The authors of [95] studied the planning decisions for siting and sizing EB charging stations. For this, an optimization charging scheduling framework was developed, which minimized the annual system operating costs, including a recharging wait cost as a penalty term. This problem was solved by a MILP and applied to a case study of the city of Davis, California, USA. Moreover, it has been demonstrated that EBs are more economical, compared to diesel buses. In the same line, in [96], the siting and sizing of fast-charging stations were proposed, considering energy consumption uncertainty. The model minimized system costs subject to electric and bus constraints. The problem was solved by a MILP with the GAMS software, and a robust optimization was applied to tackle the energy uncertainty. It was applied to the Salt Lake City (USA) bus system.

Economic planning has been another objective for some authors. In [97], the planning and operation of an EB charging station were studied, based on a Time-of-Use pricing scheme. The long-term planning model considered minimizing the energy costs, considering the operational constraints of a bus company of Taiwan and using the data from the Taiwan Power Co. The authors of [98] examined the long-term planning of three different power supply systems: wireless charging, battery swapping, and plug-in charging. The analysis was performed in Daegu City, South Korea. The results showed that investing in battery swapping stations was more beneficial than the other options for single and composite routes. Similarly, the three different power supply system costs were analyzed in [99], but considering only investment costs. The objective was to search for optimal design models by minimizing costs. Several parameters relative to EB circulation were analyzed, resulting in different investment costs for each case. In the same line, ref. [100] presented the long-term planning of EB fast-charging stations,

Considering load aggregation and renewable integration. The model optimized the investment planning and operation schedule of the EB system, based on a MILP formulation. Furthermore, an aggregation strategy was considered for co-ordination with PV resources and energy storage, considering bus route constraints. The uncertainties were tackled through Chance-constrained programming. A two-stage stochastic program for the BSS location problem to minimize the investment and operation costs was proposed in [101]. The case of Melbourne, Australia, was studied.

Another exciting approach considers the scheduling of a wirelessly charged EB system, as presented in [102]. For this, an optimal methodology that minimizes system electricity costs was developed, considering the characteristics of the WCS. The problem was solved by convex optimization, and it was demonstrated in the Guangzhou (China) Bus Rapid Transit system. Note that this methodology is considered to be a constraint to passenger satisfaction, which is associated with the passenger wait time and bus crowdedness.

By integrating large amounts of EBs into the grid, it is also crucial to maintian proper power quality levels in the network. In particular, a significant EB load leads to voltage drops. Hence, ref [103] proposed an energy management system for a smart public transportation network to regulate voltage levels. This system included solar plants and energy storage devices at EB stops. To control the energy flows, an fuzzy logic controller was used. The model was illustrated in the electrified transportation network in Guwahati City, India, showing proper voltage profiles and, thus, avoiding grid issues.

Considering the electric market transactions, the operation of EBs in the framework of Virtual Power Plants can be considered. Hence, in [104], the profit maximization of a Virtual Power Plant integrating EBs was considered by enabling the provision of grid services. Based on MILP optimization, the problem was solved considering energy procurement constraints, and proving the opportunity for EBs and Virtual Power Plants to jointly provide grid services.

The analyzed EB strategies are summarized in Table 5, and the tools used in Table 6.

Table 5. Main Objectives of EB-related Papers.

Main Objectives	References
EB Load Prediction	[85–87]
Minimized costs of EB charging stations	[88–91]
Minimized cost of EB BSSs	[92,93,101]
Siting and sizing of EB charging stations	[94–96]
Minimized EB stop time	[94]
Planning EB charging stations	[97–101]
Minimized system costs of a wireless charged EB system	[102]
Voltage regulation	[103]
Profit Maximization of Virtual Power Plant	[104]

Table 6. Tools used.

Tools Used	References
Mixed Integer Non-Linear Programming	[88]
Multi-objective optimization	[94]
MILP	[90,95,100]
Heuristic method	[90,91,105]
Genetic Algorithm Optimization	[92]
Fuzzy Logic	[103]

3.3. Approaches of Both Electric Taxis and Buses

As the operation of ETs and EBs generally require different needs, only a few studies have considered the both ETs and EBs. These works considered general aspects of their combined integration, such as the electrical load, environmental elements, and economic costs. These works are described in this section.

Some authors have considered the effect of smart charging techniques. In [106], the potential and economics of EB and ET smart charging were studied, focusing on the case of Shanghai, China. It was concluded that the possibilities for EB and ET smart charging are different and that a specific charging tariff with a high peak-valley price gap is compulsory to obtain benefits from EV smart charging. The authors of [107] proposed a charging strategy for both EBs and ETs for isolated systems with high penetration of renewable generation. Based on the proposed pricing, which maximized renewable generation use, the methodology optimized the charging load of EBs and ETs in the case study of Santa Cruz, Galapagos Islands, Ecuador. The results demonstrated that it is possible to increase the renewable generation use for EV charging, compared to uncoordinated charging. Another example of the minimization of charging cost was presented in [105], where the system load profile was also optimized and satisfied use charging requirements. The problem was solved by quadratic optimization. Case studies of Guangdong Province, China were performed, demonstrating system peak demand and charging cost reduction.

Another work which focused on both EBs and ETs dealt with BSSs. This choice must be seriously considered for EVs that require both significant electric energy consumption over a day and small time duration for charging. Therefore, the authors of [108] studied battery and charger planning for battery switch stations, minimizing investments, maintenance, and electricity costs.

Deploying ETs and EBs generates a new load that must be met by the power generation system. Thus, the question of optimal generation mix arises, considering the emissions involved. In [109], the environmental effects of ETs and EBs were considered in Beijing, indicating a reduction of millions of kilograms of CO_2 would result if they were deployed. Then, the power generation structure was analyzed. In a micro-grid case, the authors of [110] presented power generation planning considering the integration of EBs and ETs. For this purpose, the charging loads of EBs and ETs for different scenarios were modeled to obtain the optimal generation portfolio. This study was performed in the Micro-grid of Santa Cruz, Ecuador, and the results demonstrated that the best option resulted from investing in new PV generation.

4. Electric Vehicles for Public Transportation Projects over the World

Various projects have been developed around the world for electrifying public transportation. These projects have included the deployment of EVs in public transportation in cities. Some of the most relevant projects are listed as follows:

- The Green eMotion project is a European project with a budget of €42 million. It is part of the European Green Cars Initiative (EGCI) that was launched within the context of the European Recovery Plan. It is composed of 42 partners from industry, the energy sector, EV manufacturers,

municipalities, universities, and research institutions. Although this project has focused primarily on private EVs, some works have been done with ETs in Ireland [111].
- The Smart Electric Bus (BEI) project in the city of Vitoria-Gasteiz, Spain. It was the first all-electric zero emissions line of Vitoria-Gasteiz and Euskadi. This transport system was implemented by the companies Irizar e-mobility, Yarritu, and LKS, and was composed of a fleet of 13 EBs [112].
- The NEXT-E project represents a co-operation of four leading companies from the electricity and oil and gas sectors with OEMs (car manufacturers) to create an inter-operable and non-discriminatory EV charging network. The project objective was to address the issues of a continuous and cost-effective network which could allow long-distance and cross-border driving. The NEXT-E project was funded with €18.84 million to implement the project [113].
- CIVITAS is a project whose aim is to propose alternative sustainable options for transportation. It was launched by the European Commission in 2002 and funded by the European Union. The project is complemented and supported by several research and innovation projects (ECCENTRIC, PORTIS, and DESTINATIONS) [114].
- Within the ELCIDIS project, seven European cities and CITELEC co-operate, together with the European Association of cities interested in the use of EVs. This project was created to demonstrate the possibilities of a more efficient city distribution system that works with clean EVs (hybrids). The projects of the cities of Rotterdam and Stockholm focus on the deployment of large electric vans. The city project in La Rochelle focuses on deploying EVs with a payload of approximately 500 kg. The projects of the city of Stavanger, Milan, and Erlangen will focus on the deployment of hybrid EVs for the internal distribution of goods and mail for companies [115].
- In Santiago de Chile, Metbus is one of the largest pure EB fleet operators in Latin America. The project seeks to reach 3500 EBs. Currently, the projected fleet is 411 Chilean EBs, and the brands of the EBs are BYD and Yutong. EBs can accommodate up to 38 passengers each. The range, according to the manufacturer, is 250 km and the operational costs have been reduced by 76%, compared to traditional buses [116].
- The 16,359 public buses of Shenzhen are now electric. A new law passed in 2017, requires that from 2018, all buses providing transport services for passengers within Shenzhen must be electric. It is prohibited to use internal combustion buses. The measure applies to both public service buses and private buses. With this law, Shenzhen expects to reduce its emissions by up to 1.35 million tons of CO_2 per year. This is a project of 490 million dollars, which was obtained by government subsidies and the companies responsible for the assembly of the electrical infrastructure and the manufacture of buses.
- In Pengcheng, the partners have some incentives, such as a free commercial operation license for 12 years for each E-taxi of the municipality of Shenzhen. Furthermore, the Shenzhen municipality exempts customers of an E-taxi from paying a fossil fuel surcharge. The charging stations for the E-taxi fleet were jointly built by Pengcheng, BYD, and CSG [117].
- A research project named EMIL ("E-Mobility using inductive charging") in the city of Braunschweig. There are already five inductive charging stations for EB charging. The main goals are the conceptual design and the required measurements to charge ETs on the proprietary EMIL charging stations, considering the development of a system that can be applied to a variety of different vehicles [118].
- Amsterdam Airport Schiphol and the surrounding area are part of the Amstelland–Meerlanden concession, which will include 258 EBs by 2021 and has a value of around EUR 100 million per year. Schiphol Airport financed and built the infrastructure for the charging points on the Schiphol grounds. The EBs operate 24/7 and have a battery capacity of 170 kWh [119].
- The purpose of the TransLink demonstration project is to guide the installation and operation of overhead charging stations for EBs in Vancouver. The Government of Canada has encouraged the widespread adoption of EVs by supporting projects which provide more options for sustainable transportation, demonstrating new and innovative charging technologies. The Government has invested $182.5 million to support the deployment of electric chargers, natural gas and hydrogen

refueling stations, and the development of standards. New Flyer Industries and Nova Bus are developing electric transit buses, while ABB and Siemens are developing the chargers. The project to evaluate inter-operability and performance will integrate more than one bus manufacturer and more than one charging system provider [120].

- The capital of Ukraine, Kiev, has a project where electric mobility is emerging as a solution for service vehicles, approximately 30 Nissan Leafs. On average, each car travels about 100 km per load and requires four load cycles per day. The company that owns the vehicles has reached an agreement with a fast-charging infrastructure operator and serves approximately 20 locations in Kiev, and each driver has a smart card that is used to pay at charging station [121].
- In Poland, the implementation of a national e-mobility strategy, including electric public transport and the related infrastructure, is currently backed by approximately €2.3 billion worth of governmental financial incentives for the period 2018–2028, mainly supported by the Low-Emission Transport Fund, and the Emission-Free Public Transport (BTP) program, as well as the European Structural and Investment Funds. Currently, around 189 EBs are operating in Polish cities, with an additional 390 vehicles that will be incorporated by the end of 2019 [122].

5. Future Trends

This section provides some insights for future research works related to the impact of ETs and EBs in the smart grid. As mentioned above, the number of works in this area is relatively small and recent, so future outlooks are described.

5.1. ET and EB Aggregator

The EV aggregator is a new player in the electricity market, which collects EVs by attracting and retaining them as a significant load that can beneficially impact the grid. The size of aggregation is crucial to ensure its proper role. An EV aggregator could work as a controllable load or as a resource [15].

The works considering EB or ET aggregators are few. Only the works [103,104] proposed an ET aggregator and the works [78,79] an EB aggregator. Furthermore, it should be noted that in various works, the BSS has been considered to be an entity that works as an aggregator, considering that it aggregates all batteries that are charged daily, and could offer services for the grid.

5.2. Micro-Grids with RES and ETs or EBs

Several works have considered using EVs as support for the integration of RES into the grid. However, in the main grids, RES are already integrated into the grid constraints and electricity markets. One interesting topic that has been increasing in popularity is the study of the operation of micro-grids with RES.

A micro-grid is described as a cluster of loads, distributed generation units, and energy storage systems, operated in co-ordination to reliably supply electricity, either connected to a host power system at the distribution level at a single point of connection or in isolation from the bulk grid. Moreover, a micro-grid can operate in grid-connected and stand-alone modes, depending on grid conditions and the transitions between these two modes [123].

Some remote communities or islands located far away from a mainland depend on isolated micro-grids based on diesel fuel, which is environmentally harmful [124]. These effects are even worse if pristine and protected environments are considered [110]. Therefore, micro-grids with RES are being developed, creating concerns for micro-grid stability due to the RES generation uncertainties and fluctuations. ETs and EBs seem to be a suitable solution for providing clean transportation, as well as to mitigate micro-grid stability issues with the use of their batteries as energy storage.

Only [107,110] have addressed the impact of electric public transportation on a micro-grid with RES. It should be that expected more future works will emerge in this field.

5.3. V2G for EBS and ETS

V2G is considered to be a technique where EVs do not only absorb electricity from the grid, but can also supply electricity to the grid from their batteries, when the grid requires it [14]. V2G can offer four different services to the grid: baseload power, peak power, spinning reserves, and ancillary services. The last two have been considered to be ancillary services [14].

This trend has widely been studied for private EVs. As EBs and ETs possess higher-capacity batteries, V2G could be implemented for BSSs. In particular, BSSs could have time flexibility concerning WSS and plug-in fast charging.

Only [73] have considered V2G for an ET BSS. No work has considered V2G for EBs BSS. Therefore, this could be an insight for future research.

6. Conclusions

EVs will play a key role in the future smart grid, and are necessary for the reduction of polluting gases in cities. In particular, more attention has been devoted in the past few years to the introduction of EVs for public transportation. To achieve an adequate mass introduction of these EVs, it is essential to propose approaches that mitigate grid issues and improve power systems stability. Therefore, various methodologies have been established in the literature recently. In this paper, a review of different approaches for the integration of EVs for public transportation into power systems has been presented.

First, a brief background of the typical models used for ETs and EBs was presented. In particular, EVs for public transportation generally possess batteries with higher battery capacities than typical private EVs, and a bigger distance range. The four different types of charging were, then, introduced (plug-in charging, WCS, BSS, and pantograph).

The various works covered a broad range of objectives for ETs and EBs, such as siting and sizing charging stations, cost minimization, load unbalance minimization, planning of charging stations, and so on. Many of these works used optimization methods to solve their problems.

Several governments and organizations around the world are promoting the adoption of EVs by deploying pilot studies for the use of ETs and EBs, which were presented in this review.

Finally, this paper provides some insights for future research in this area, such as the interaction of ETs and EBs with aggregators, their integration into micro-grids with RES, and their participation in V2G services.

Author Contributions: Conceptualization, J.-M.C.; Data curation, J.-M.C., P.G.-T., and X.S.-G.; Formal analysis, J.-M.C.; Investigation, J.-M.C., P.G.-T., and X.S.-G.; Methodology, J.-M.C.; Supervision, M.G.-R. and G.E.-E.; Validation, M.G.-R. and G.E.-E.; Writing—original draft, J.-M.C., P.G.-T., and X.S.-G.; Writing—review and editing, J.-M.C., M.G.-R., and G.E.-E.

Funding: This research was funded by the project SIS.JCG.19.03 of Universidad de las Américas, Ecuador.

Conflicts of Interest: The authors declare no conflict of interest.

Abbreviations

The following abbreviations are used in this manuscript:

BSS	Battery Swapping Station
EB	Electric Bus
ET	Electric Taxi
EV	Electric Vehicle
ICV	Internal Combustion Vehicle
MILP	Mixed Integer Linear Programming
PV	Photovoltaic
RES	Renewable Energy Source
V2G	Vehicle-to-Grid
WCS	Wireless Charging System

References

1. Emadi, A. Transportation 2.0. *IEEE Power Energy Mag.* **2011**, *9*, 54–64. [CrossRef]
2. Fahimi, B.; Kwasinski, A.; Davoudi, A.; Balog, R.S.; Kiani, M. Charge It! *IEEE Power Energy Mag.* **2011**, *9*, 54–64. [CrossRef]
3. Yilmaz, M.; Krein, P.T. Review of charging power levels and infrastructure for plug-in electric and hybrid vehicles. *IEEE Trans. Power Electron.* **2012**, *28*, 2151–2169. [CrossRef]
4. Tagliaferri, C.; Evangelisti, S.; Acconcia, F.; Domenech, T.; Ekins, P.; Barletta, D.; Lettieri, P. Life cycle assessment of future electric and hybrid vehicles: A cradle-to-grave systems engineering approach. *Chem. Eng. Res. Des.* **2016**, *112*, 298–309. [CrossRef]
5. Zackrisson, M.; Fransson, K.; Hildenbrand, J.; Lampic, G.; O'Dwyer, C. Life cycle assessment of lithium-air battery cells. *J. Clean. Prod.* **2016**, *135*, 299–311. [CrossRef]
6. Wu, Y.; Yang, Z.; Lin, B.; Liu, H.; Wang, R.; Zhou, B.; Hao, J. Energy consumption and CO_2 emission impacts of vehicle electrification in three developed regions of China. *Energy Policy* **2012**, *48*, 537–550. [CrossRef]
7. Shen, W.; Han, W.; Chock, D.; Chai, Q.; Zhang, A. Well-to-wheels life-cycle analysis of alternative fuels and vehicle technologies in China. *Energy Policy* **2012**, *49*, 296–307. [CrossRef]
8. Wang, R.; Wu, Y.; Ke, W.; Zhang, S.; Zhou, B.; Hao, J. Can propulsion and fuel diversity for the bus fleet achieve the win-win strategy of energy conservation and environmental protection? *Appl. Energy* **2015**, *147*, 92–103. [CrossRef]
9. Clement-Nyns, K.; Haesen, E.; Driesen, J. The impact of Charging plug-in hybrid electric vehicles on a residential distribution grid. *IEEE Trans. Power Syst.* **2009**, *25*, 371–380. [CrossRef]
10. Shafiee, S.; Fotuhi-Firuzabad, M.; Rastegar, M. Investigating the impacts of plug-in hybrid electric vehicles on power distribution systems. *IEEE Trans. Smart Grid* **2013**, *4*, 1351–1360. [CrossRef]
11. Pieltain Fernández, L.; Gómez San Román, T.; Cossent, R.; Mateo Domingo, C.; Frías, P. Assessment of the impact of plug-in electric vehicles on distribution networks. *IEEE Trans. Power Syst.* **2011**, *26*, 206–213. [CrossRef]
12. Lucas, A.; Bonavitacola, F.; Kotsakis, E.; Fulli, G. Grid harmonic impact of multiple electric vehicle fast charging. *Electr. Power Syst. Res.* **2015**, *127*, 13–21. [CrossRef]
13. Turker, H.; Bacha, S.; Chatroux, D.; Hably, A. Low-voltage transformer loss-of-life assessments for a high penetration of plug-in hybrid electric vehicles (PHEVs). *IEEE Trans. Power Deliv.* **2012**, *27*, 1323–1331. [CrossRef]
14. Kempton, W.; Tomić, J. Vehicle-to-grid power fundamentals: Calculating capacity and net revenue. *J. Power Sources* **2005**, *144*, 268–279. [CrossRef]
15. Guille, C.; Gross, G. A conceptual framework for the vehicle-to-grid (V2G) implementation. *Energy Policy* **2009**, *37*, 4379–4390. [CrossRef]
16. Geng, Z.; Conejo, A.J.; Chen, Q.; Xia, Q.; Kang, C. Electricity production scheduling under uncertainty: Max social welfare vs. min emission vs. max renewable production. *Appl. Energy* **2017**, *193*, 540–549. [CrossRef]
17. Verbruggen, A.; Fischedick, M.; Moomaw, W.; Weir, T.; Nadaï, A.; Nilsson, L.J.; Nyboer, J.; Sathaye, J. Renewable energy costs, potentials, barriers: Conceptual issues. *Energy Policy* **2010**, *38*, 850–861. [CrossRef]
18. Oda, T.; Aziz, M.; Mitani, T.; Watanabe, Y.; Kashiwagi, T. Mitigation of congestion related to quick charging of electric vehicles based on waiting time and cost–benefit analyses: A japanese case study. *Sustain. Cities Soc.* **2018**, *36*, 99–106. [CrossRef]
19. Arkin, E.M.; Carmi, P.; Katz, M.J.; Mitchell, J.S.; Segal, M. Locating battery charging stations to facilitate almost shortest paths. *Discret. Appl. Math.* **2019**, *254*, 10–16. [CrossRef]
20. Gallardo-Lozano, J.; Milanés-Montero, M.I.; Guerrero-Martínez, M.A.; Romero-Cadaval, E. Electric vehicle battery charger for smart grids. *Electr. Power Syst. Res.* **2012**, *90*, 18–29. [CrossRef]
21. Aziz, M.; Oda, T.; Ito, M. Battery-assisted charging system for simultaneous charging of electric vehicles. *Energy* **2016**, *100*, 82–90. [CrossRef]
22. Mehboob, N.; Restrepo, M.; Canizares, C.A.; Rosenberg, C.; Kazerani, M. Smart Operation of Electric Vehicles with Four-Quadrant Chargers Considering Uncertainties. *IEEE Trans. Smart Grid* **2018**, *10*, 2999–3009. [CrossRef]

23. García-Villalobos, J.; Zamora, I.; San Martín, J.I.; Asensio, F.J.; Aperribay, V. Plug-in electric vehicles in electric distribution networks: A review of smart charging approaches. *Renew. Sustain. Energy Rev.* **2014**, *38*, 717–731. [CrossRef]
24. Richardson, D.B. Electric vehicles and the electric grid: A review of modeling approaches, Impacts, and renewable energy integration. *Renew. Sustain. Energy Rev.* **2013**, *19*, 247–254. [CrossRef]
25. Haidar, A.M.; Muttaqi, K.M.; Sutanto, D. Technical challenges for electric power industries due to grid-integrated electric vehicles in low voltage distributions: A review. *Energy Convers. Manag.* **2014**, *86*, 689–700. [CrossRef]
26. Mwasilu, F.; Justo, J.J.; Kim, E.K.; Do, T.D.; Jung, J.W. Electric vehicles and smart grid interaction: A review on vehicle to grid and renewable energy sources integration. *Renew. Sustain. Energy Rev.* **2014**, *34*, 501–516. [CrossRef]
27. Habib, S.; Kamran, M.; Rashid, U. Impact analysis of vehicle-to-grid technology and charging strategies of electric vehicles on distribution networks—A review. *J. Power Sources* **2015**, *277*, 205–214. [CrossRef]
28. Tan, K.M.; Ramachandaramurthy, V.K.; Yong, J.Y. Integration of electric vehicles in smart grid: A review on vehicle to grid technologies and optimization techniques. *Renew. Sustain. Energy Rev.* **2016**, *53*, 720–732. [CrossRef]
29. Raslavičius, L.; Azzopardi, B.; Keršys, A.; Starevičius, M.; Bazaras, Ž.; Makaras, R. Electric vehicles challenges and opportunities: Lithuanian review. *Renew. Sustain. Energy Rev.* **2015**, *42*, 786–800. [CrossRef]
30. Rahman, I.; Vasant, P.M.; Singh, B.S.M.; Abdullah-Al-Wadud, M.; Adnan, N. Review of recent trends in optimization techniques for plug-in hybrid, and electric vehicle charging infrastructures. *Renew. Sustain. Energy Rev.* **2016**, *58*, 1039–1047. [CrossRef]
31. Faddel, S.; Al-Awami, A.T.; Mohammed, O.A. Charge control and operation of electric vehicles in power grids: A review. *Energies* **2018**, *11*, 701. [CrossRef]
32. Vuchic, V.R. *Urban Transit Systems and Technology*; John Wiley & Sons: Hoboken, NJ, USA, 2007.
33. Ercan, T.; Onat, N.C.; Tatari, O. Investigating carbon footprint reduction potential of public transportation in United States: A system dynamics approach. *J. Clean. Prod.* **2016**, *133*, 1260–1276. [CrossRef]
34. Kwan, S.C.; Hashim, J.H. A review on co-benefits of mass public transportation in climate change mitigation. *Sustain. Cities Soc.* **2016**, *22*, 11–18. [CrossRef]
35. Kolbe, K. Mitigating urban heat island effect and carbon dioxide emissions through different mobility concepts: Comparison of conventional vehicles with electric vehicles, hydrogen vehicles and public transportation. *Transp. Policy* **2019**, *80*, 1–11. [CrossRef]
36. Zalakeviciute, R.; Rybarczyk, Y.; López-Villada, J.; Diaz Suarez, M.V. Quantifying decade-long effects of fuel and traffic regulations on urban ambient PM2.5 pollution in a mid-size South American city. *Atmos. Pollut. Res.* **2018**, *9*, 66–75. [CrossRef]
37. Dell'Olio, L.; Ibeas, A.; Cecin, P. The quality of service desired by public transport users. *Transp. Policy* **2011**, *18*, 217–227. [CrossRef]
38. Mahmoud, M.; Garnett, R.; Ferguson, M.; Kanaroglou, P. Electric buses: A review of alternative powertrains. *Renew. Sustain. Energy Rev.* **2016**, *62*, 673–684. [CrossRef]
39. Nissan. Nissan Leaf. Available online: https://www.nissan.co.uk/vehicles/new-vehicles/leaf/range-charging.html (accessed on 30 June 2019).
40. Kia. Introducing the Fully Charged 2020 Kia Soul EV. Available online: https://www.kia.com/us/en/content/vehicles/upcoming-vehicles/2020-soul-ev (accessed on 30 June 2019).
41. BYD. e6. Available online: https://en.byd.com/wp-content/uploads/2017/06/e6_cutsheet.pdf (accessed on 30 June 2019).
42. Tesla. Tesla Model S. Available online: https://www.tesla.com/models (accessed on 30 June 2019).
43. BYD. Bus. Available online: https://en.byd.com/bus/40-electric-motor-coach/ (accessed on 30 June 2019).
44. Solaris. Urbino Electric. 2019. Available online: https://www.solarisbus.com/en/vehicles/zero-emissions/urbino-electric (accessed on 24 July 2019).
45. Volvo. Volvo 7900 Electric. Available online: https://www.volvobuses.co.uk/en-gb/our-offering/buses/volvo-7900-electric/specifications.html (accessed on 24 July 2019).
46. Collin, R.; Miao, Y.; Yokochi, A.; Enjeti, P.; von Jouanne, A. Advanced Electric Vehicle Fast-Charging Technologies. *Energies* **2019**, *12*, 1839. [CrossRef]

47. Yang, Y.; El Baghdadi, M.; Lan, Y.; Benomar, Y.; Van Mierlo, J.; Hegazy, O. Design methodology, modeling, and comparative study of wireless power transfer systems for electric vehicles. *Energies* **2018**, *11*, 1716. [CrossRef]
48. Bi, Z.; Song, L.; De Kleine, R.; Mi, C.C.; Keoleian, G.A. Plug-in vs. wireless charging: Life cycle energy and greenhouse gas emissions for an electric bus system. *Appl. Energy* **2015**, *146*, 11–19. [CrossRef]
49. Li, S.; Mi, C.C. Wireless power transfer for electric vehicle applications. *IEEE J. Emerg. Sel. Top. Power Electron.* **2015**, *3*, 4–17. [CrossRef]
50. Eberle, W.; Musavi, F. Overview of wireless power transfer technologies for electric vehicle battery charging. *IET Power Electron.* **2013**, *7*, 60–66. [CrossRef]
51. Wang, Z.; Wei, X.; Dai, H. Design and control of a 3 kW wireless power transfer system for electric vehicles. *Energies* **2016**, *9*, 10. [CrossRef]
52. Sarker, M.R.; Pandžić, H.; Ortega-Vazquez, M.A. Optimal operation and services scheduling for an electric vehicle battery swapping station. *IEEE Trans. Power Syst.* **2015**, *30*, 901–910. [CrossRef]
53. Adegbohun, F.; von Jouanne, A.; Lee, K.Y. Autonomous battery swapping system and methodologies of electric vehicles. *Energies* **2019**, *12*, 667. [CrossRef]
54. OPPCharge. OPPChargeCommon Interface for Automated Charging of Hybrid Electric and Electric Commercial Vehicles. 2019. Available online: https://www.oppcharge.org/dok/OPPChargeSpecification2ndedition20190421.pdf (accessed on 24 July 2019).
55. OPPCharge. Fast Charging of Electric Vehicles. 2019. Available online: https://www.oppcharge.org (accessed on 24 July 2019).
56. Ruoff, C. The inevitability of electric buses. *Charged* **2016**, 40–51.
57. Jiang, C.X.; Jing, Z.X.; Cui, X.R.; Ji, T.Y.; Wu, Q.H. Multiple agents and reinforcement learning for modelling charging loads of electric taxis. *Appl. Energy* **2018**, *222*, 158–168. [CrossRef]
58. Fraile-Ardanuy, J.; Castano-Solis, S.; Álvaro-Hermana, R.; Merino, J.; Castillo, Á. Using mobility information to perform a feasibility study and the evaluation of spatio-temporal energy demanded by an electric taxi fleet. *Energy Convers. Manag.* **2018**, *157*, 59–70. [CrossRef]
59. Rao, R.; Cai, H.; Xu, M. Modeling electric taxis' charging behavior using real-world data. *Int. J. Sustain. Transp.* **2018**, *12*, 452–460. [CrossRef]
60. He, Z.; Cheng, Y.; Hu, Z. Multi-Time Simulation of Electric Taxicabs' Charging Demand Based on Residents' Travel Characteristics. In Proceedings of the 2017 IEEE Conference on Energy Internet and Energy System Integration (EI2), Beijing, China, 26–28 November 2017; pp. 5–10.
61. Litzlbauer, M. Technische Machbarkeitsanalyse einer rein elektrisch betriebenen TaxiflotteTechnical feasibility study of a purely electrically driven taxi fleet. *Elektrotech. Inf.* **2015**, *132*, 172–177. [CrossRef]
62. Liao, B.; Li, L.; Li, B.; Mao, J.; Yang, J.; Wen, F.; Salam, M.A. Load modeling for electric taxi battery charging and swapping stations: Comparison studies. In Proceedings of the 2016 IEEE 2nd Annual Southern Power Electronics Conference (SPEC), Auckland, New Zealand, 5–8 December 2016; p. 150DUMMY. [CrossRef]
63. Zou, Y.; Wei, S.; Sun, F.; Hu, X.; Shiao, Y. Large-scale deployment of electric taxis in Beijing: A real-world analysis. *Energy* **2016**, *100*, 25–39. [CrossRef]
64. Asamer, J.; Reinthaler, M.; Ruthmair, M.; Straub, M.; Puchinger, J. Optimizing charging station locations for urban taxi providers. *Transp. Res. Part A Policy Pract.* **2016**, *85*, 233–246. [CrossRef]
65. Yang, J.; Dong, J.; Hu, L. A data-driven optimization-based approach for siting and sizing of electric taxi charging stations. *Transp. Res. Part C Emerg. Technol.* **2017**, *77*, 462–477. [CrossRef]
66. Jiang, C.; Jing, Z.; Ji, T.; Wu, Q. Optimal location of PEVCSs using MAS and ER approach. *IET Gener. Transm. Distrib.* **2018**, *12*, 4377–4387. [CrossRef]
67. Pan, A.; Zhao, T.; Yu, H.; Zhang, Y. Deploying Public Charging Stations for Electric Taxis: A Charging Demand Simulation Embedded Approach. *IEEE Access* **2019**, *7*, 17412–17424. [CrossRef]
68. Lianfu, C.; Zhang, W.; Huang, Y.; Zhang, D. Research on the charging station service radius of electric taxis. In Proceedings of the 2014 IEEE Conference and Expo Transportation Electrification Asia-Pacific (ITEC Asia-Pacific), Beijing, China, 31 August–3 September 2014; pp. 1–4. [CrossRef]
69. Yang, Y.; Zhang, W.; Niu, L.; Jiang, J. Coordinated Charging Strategy for Electric Taxis in Temporal and Spatial Scale. *Energies* **2015**, *8*, 1256–1272. [CrossRef]
70. Niu, L.; Zhang, D. Charging Guidance of Electric Taxis Based on Adaptive Particle Swarm Optimization. *Sci. World J.* **2015**, *2015*, 354952. [CrossRef]

71. Yang, Z.; Guo, T.; You, P.; Hou, Y.; Qin, S.J. Distributed Approach for Temporal-Spatial Charging Coordination of Plug-in Electric Taxi Fleet. *IEEE Trans. Ind. Inform.* **2018**, *15*, 3185–3195. [CrossRef]
72. Rossi, F.; Iglesias, R.D.; Alizadeh, M.; Pavone, M. On the interaction between Autonomous Mobility-on-Demand systems and the power network: Models and coordination algorithms. *IEEE Trans. Control Netw. Syst.* **2019**, *5870*, 1–12. [CrossRef]
73. Liang, Y.; Zhang, X.; Xie, J.; Liu, W. An Optimal Operation Model and Ordered Charging/Discharging Strategy for Battery Swapping Stations. *Sustainability* **2017**, *9*, 700. [CrossRef]
74. Xu, X.; Yao, L.; Zeng, P. Architecture and performance analysis of a smart battery charging and swapping operation service network for electric vehicles in China. *J. Mod. Power Syst. Clean Energy* **2015**, *3*, 259–268. [CrossRef]
75. Jing, Z.; Fang, L.; Lin, S.; Shao, W. Modeling for electric taxi load and optimization model for charging/swapping facilities of electric taxi. In Proceedings of the 2014 IEEE Conference and Expo Transportation Electrification Asia-Pacific (ITEC Asia-Pacific), Beijing, China, 31 August–3 September 2014; pp. 1–5. [CrossRef]
76. Wang, Y.; Ding, W.; Huang, L.; Wei, Z.; Liu, H.; Stankovic, J.A. Toward Urban Electric Taxi Systems in Smart Cities: The Battery Swapping Challenge. *IEEE Trans. Veh. Technol.* **2018**, *67*, 1946–1960. [CrossRef]
77. You, P.; Yang, Z.; Zhang, Y.; Low, S.H.; Sun, Y. Optimal Charging Schedule for a Battery Switching Station Serving Electric Buses. *IEEE Trans. Power Syst.* **2016**, *31*, 3473–3483. [CrossRef]
78. Yang, Z.; Sun, L.; Chen, J.; Yang, Q.; Chen, X.; Xing, K. Profit maximization for plug-in electric taxi with uncertain future electricity prices. *IEEE Trans. Power Syst.* **2014**, *29*, 3058–3068. [CrossRef]
79. Yang, Z.; Sun, L.; Ke, M.; Shi, Z.; Chen, J. Optimal charging strategy for plug-in electric taxi with time-varying profits. *IEEE Trans. Smart Grid* **2014**, *5*, 2787–2797. [CrossRef]
80. Yang, J.; Xu, Y.; Yang, Z. Regulating the Collective Charging Load of Electric Taxi Fleet via Real-Time Pricing. *IEEE Trans. Power Syst.* **2017**, *32*, 3694–3703. [CrossRef]
81. Du, R.; Liao, G.; Zhang, E.; Wang, J. Battery charge or change, which is better? A case from Beijing, China. *J. Clean. Prod.* **2018**, *192*, 698–711. [CrossRef]
82. Chen, H.; Jia, Y.; Hu, Z.; Wu, G.; Shen, Z.J.M. Data-driven planning of plug-in hybrid electric taxi charging stations in urban environments: A case in the central area of Beijing. In Proceedings of the 2017 IEEE PES Innovative Smart Grid Technologies Conference Europe (ISGT-Europe), Torino, Italy, 26–29 September 2017.
83. Yang, J.; Dong, J.; Lin, Z.; Hu, L. Predicting market potential and environmental benefits of deploying electric taxis in Nanjing, China. *Transp. Res. Part D Transp. Environ.* **2016**, *49*, 68–81. [CrossRef]
84. You, P.; Low, S.H.; Yang, Z.; Zhang, Y.; Fu, L. Real-time recommendation algorithm of battery swapping stations for electric taxis. In Proceedings of the 2016 IEEE Power and Energy Society General Meeting (PESGM), Boston, MA, USA, 17–21 July 2016; pp. 1–5. [CrossRef]
85. Dai, Q.; Cai, T.; Duan, S.; Zhao, F. Stochastic modeling and forecasting of load demand for electric bus battery-swap station. *IEEE Trans. Power Deliv.* **2014**, *29*, 1909–1917. [CrossRef]
86. Mohamed, M.; Farag, H.; El-Taweel, N.; Ferguson, M. Simulation of electric buses on a full transit network: Operational feasibility and grid impact analysis. *Electr. Power Syst. Res.* **2017**, *142*, 163–175. [CrossRef]
87. Zhang, X. Short-term load forecasting for electric bus charging stations based on fuzzy clustering and least squares support vector machine optimized by Wolf pack algorithm. *Energies* **2018**, *11*, 1449. [CrossRef]
88. Ding, H.; Hu, Z.; Song, Y. Value of the energy storage system in an electric bus fast charging station. *Appl. Energy* **2015**, *157*, 630–639. [CrossRef]
89. Qin, N.; Gusrialdi, A.; Paul Brooker, R.; T-Raissi, A. Numerical analysis of electric bus fast charging strategies for demand charge reduction. *Transp. Res. Part A Policy Pract.* **2016**, *94*, 386–396. [CrossRef]
90. Chen, H.; Hu, Z.; Xu, Z.; Li, J.; Zhang, H.; Xia, X.; Ning, K.; Peng, M. Coordinated charging strategies for electric bus fast charging stations. In Proceedings of the 2016 IEEE PES Asia-Pacific Power and Energy Engineering Conference (APPEEC), Xi'an, China, 25–28 October 2016; pp. 1174–1179. [CrossRef]
91. Chen, H.; Hu, Z.; Zhang, H.; Luo, H. Coordinated charging and discharging strategies for plug-in electric bus fast charging station with energy storage system. *IET Gener. Transm. Distrib.* **2018**, *12*, 2019–2028. [CrossRef]
92. Gao, Y.; Guo, S.; Ren, J.; Zao, Z.; Ehsan, A.; Zheng, Y. An Electric Bus Power Consumption Model and Optimization of Charging Scheduling Concerning Multi-External Factors. *Energies* **2018**, *11*, 2060. [CrossRef]

93. Cheng, Y.; Tao, J. Optimization of A Micro Energy Network Integrated with Electric Bus Battery Swapping Station and Distributed PV. In Proceedings of the 2018 2nd IEEE Conference on Energy Internet and Energy System Integration (EI2), Beijing, China, 20–22 October 2018; pp. 1–6. [CrossRef]
94. Sebastiani, M.T.; Luders, R.; Fonseca, K.V.O. Evaluating Electric Bus Operation for a Real-World BRT Public Transportation Using Simulation Optimization. *IEEE Trans. Intell. Transp. Syst.* **2016**, *17*, 2777–2786. [CrossRef]
95. Wang, Y.; Huang, Y.; Xu, J.; Barclay, N. Optimal recharging scheduling for urban electric buses: A case study in Davis. *Transp. Res. Part E Logist. Transp. Rev.* **2017**, *100*, 115–132. [CrossRef]
96. Liu, Z.; Song, Z.; He, Y. Planning of Fast-Charging Stations for a Battery Electric Bus System under Energy Consumption Uncertainty. *Transp. Res. Rec.* **2018**. [CrossRef]
97. Leou, R.C.; Hung, J.J. Optimal charging schedule planning and economic analysis for electric bus charging stations. *Energies* **2017**, *10*, 483. [CrossRef]
98. Bak, D.B.; Bak, J.S.; Kim, S.Y. Strategies for implementing public service electric bus lines by charging type in Daegu Metropolitan City, South Korea. *Sustainability* **2018**, *10*, 3386. [CrossRef]
99. Chen, Z.; Yin, Y.; Song, Z. A cost-competitiveness analysis of charging infrastructure for electric bus operations. *Transp. Res. Part C Emerg. Technol.* **2018**, *93*, 351–366. [CrossRef]
100. Cheng, Y.; Wang, W.; Ding, Z.; He, Z. Electric bus fast charging station resource planning considering load aggregation and renewable integration. *IET Renew. Power Gener.* **2019**, *13*, 1132–1141. [CrossRef]
101. An, K.; Jing, W.; Kim, I. Battery-swapping facility planning for electric buses with local charging systems. *Int. J. Sustain. Transp.* **2019**, 1–13. [CrossRef]
102. Yang, C.; Lou, W.; Yao, J.; Xie, S. On Charging Scheduling Optimization for a Wirelessly Charged Electric Bus System. *IEEE Trans. Intell. Transp. Syst.* **2018**, *19*, 1814–1826. [CrossRef]
103. Bhaskar Naik, M.; Kumar, P.; Majhi, S. Smart public transportation network expansion and its interaction with the grid. *Int. J. Electr. Power Energy Syst.* **2019**, *105*, 365–380. [CrossRef]
104. Raab, A.F.; Lauth, E.; Strunz, K.; Göhlich, D. Implementation schemes for electric bus fleets at depots with optimized energy procurements in virtual power plant operations. *World Electr. Veh. J.* **2019**, *10*, 5. [CrossRef]
105. Xu, Z.; Su, W.; Hu, Z.; Song, Y.; Zhang, H. A hierarchical framework for coordinated charging of plug-in electric vehicles in China. *IEEE Trans. Smart Grid* **2016**, *7*, 428–438. [CrossRef]
106. Jian, L.; Yongqiang, Z.; Hyoungmi, K. The potential and economics of EV smart charging: A case study in Shanghai. *Energy Policy* **2018**, *119*, 206–214. [CrossRef]
107. Clairand, J.M.; Rodríguez-García, J.; Álvarez-Bel, C. Electric Vehicle Charging Strategy for Isolated Systems with High Penetration of Renewable Generation. *Energies* **2018**, *11*, 3188. [CrossRef]
108. Lin, N.; Lin, X.; Chen, Q.; Zou, F.; Chen, Z. Optimal Configuration for Batteries and Chargers in Battery Switch Station Considering Extra Waiting Time of Electric Vehicles. *J. Energy Eng.* **2016**, *143*, 04016035. [CrossRef]
109. Ma, Y.; Ke, R.Y.; Han, R.; Tang, B.J. The analysis of the battery electric vehicle's potentiality of environmental effect: A case study of Beijing from 2016 to 2020. *J. Clean. Prod.* **2017**, *145*, 395–406. [CrossRef]
110. Clairand, J.M.; Arriaga, M.; Cañizares, C.; Alvarez-bel, C. Power Generation Planning of Galapagos' Microgrid Considering Electric Vehicles and Induction Stoves. *IEEE Trans. Sustain. Energy* **2018**, 1–12. [CrossRef]
111. GreeneMotion. Green eMotion. Available online: http://www.greenemotion-project.eu/home/home.php (accessed on 28 June 2019).
112. Econoticias. El Bus Eléctrico Inteligente que funcionará en Vitoria a partir de 2020. Available online: https://www.ecoticias.com/motor/195107/Bus-Electrico-Inteligente-funcionara-Vitoria-partir-2020 (accessed on 28 June 2019).
113. Next-e. Available online: https://next-e.eu/about.html (accessed on 28 June 2019).
114. Civitas. CIVITAS: Cleaner and Better Transport in Cities. Available online: https://civitas.eu/ (accessed on 28 June 2019).
115. Elcidis. The Elcidis Project. Available online: https://www.elcidis.org/project.htm (accessed on 28 June 2019).
116. Revista Colectibondi. Las principales flotas y proyectos con buses eléctricos alrededor del mundo. Available online: http://www.revistacolectibondi.com.ar/2019/04/21/las-principales-flotas-y-proyectos-con-buses-electricos-alrededor-del-mundo/ (accessed on 28 June 2019).

117. Li, Y.; Zhan, C.; de Jong, M.; Lukszo, Z. Business innovation and government regulation for the promotion of electric vehicle use: Lessons from Shenzhen, China. *J. Clean. Prod.* **2016**, *134*, 371–383. [CrossRef]
118. Henke, M.; Dietrich, T.H. High power inductive charging system for an electric taxi vehicle. In Proceedings of the 2017 IEEE Transportation Electrification Conference and Expo (ITEC), Chicago, IL, USA, 22–24 June 2017; pp. 27–32. [CrossRef]
119. Schiphol. Europe's Largest Fleet. Available online: https://www.schiphol.nl/en/schiphol-group/page/europes-largest-fleet-of-fully-electric-buses/ (accessed on 28 June 2019).
120. Government of Canada. Electric Bus Infrastructure Comes to Vancouver. Available online: https://www.canada.ca/en/natural-resources-canada/news/2018/04/electric-bus-infrastructure-comes-to-vancouver.html (accessed on 28 June 2019).
121. Destinations. Green Electric Cars in Oxy-Taxi Service in Kiev. Available online: https://destinations.com.ua/cars-boats/green-electric-cars-in-oxy-taxi-service-in-kiev (accessed on 28 June 2019).
122. Bayer. Electric Mobility—Materials. Available online: http://www.research.bayer.com/en/23-electric-mobility.pdf (accessed on 28 June 2019).
123. Olivares, D.E.; Mehrizi-Sani, A.; Etemadi, A.H.; Cañizares, C.A.; Iravani, R.; Kazerani, M.; Hajimiragha, A.H.; Gomis-Bellmunt, O.; Saeedifard, M.; Palma-Behnke, R.; et al. Trends in microgrid control. *IEEE Trans. Smart Grid* **2014**, *5*, 1905–1919. [CrossRef]
124. Arriaga, M.; Cañizares, C.A.; Kazerani, M. Long-Term Renewable Energy Planning Model for Remote Communities. *IEEE Trans. Sustain. Energy* **2016**, *7*, 221–231. [CrossRef]

© 2019 by the authors. Licensee MDPI, Basel, Switzerland. This article is an open access article distributed under the terms and conditions of the Creative Commons Attribution (CC BY) license (http://creativecommons.org/licenses/by/4.0/).

MDPI
St. Alban-Anlage 66
4052 Basel
Switzerland
Tel. +41 61 683 77 34
Fax +41 61 302 89 18
www.mdpi.com

Energies Editorial Office
E-mail: energies@mdpi.com
www.mdpi.com/journal/energies